Operations Research/Computer Science Interfaces Series

Volume 51

For further volumes:
http://www.springer.com/series/6375

Stephan Meisel

Anticipatory Optimization for Dynamic Decision Making

Springer

Stephan Meisel
Carl-Friedrich Gauss Department
University of Braunschweig
Mühlenpfordtstrasse 23
38106 Braunschweig
Germany
stephan.meisel@tu-braunschweig.de

ISSN 1387-666X
ISBN 978-1-4614-0504-7 e-ISBN 978-1-4614-0505-4
DOI 10.1007/978-1-4614-0505-4
Springer New York Dordrecht Heidelberg London

Library of Congress Control Number: 2011931714

Printed on acid-free paper

Springer is part of Springer Science+Business Media (www.springer.com)

To my parents Renate and Karl-Heinz

Preface

Anticipatory optimization for dynamic decision making relies on a number of different scientific disciplines. On a general level, the foundations of the field may be localized at the intersection of operations research, computer science and decision theory. Closer inspection reveals the important role of branches such as simulation, metaheuristics, Markov decision processes, dynamic programming and data mining. Moreover, realization of an advanced approach to anticipatory optimization is hardly possible without supporting database technology and without profound knowledge about the problem domain under consideration.

However, bringing together all these elements is not an end in itself. Rather the integral nature of anticipatory optimization reflects the complexity of dynamic decision problems as they increasingly occur within various operational contexts. Ultimately, the need for anticipatory optimization results from the fact that today many companys' operations are carried out under strongly dynamic and uncertain conditions.

Starting from a discussion of origins and impacts of these conditions, the following chapters elaborate on anticipatory optimization for dynamic decision making. The presentation has three major facets. First of all, it focuses on consistent integration of the methodological building blocks required for making an anticipatory decision. Moreover, the presentation distinguishes between different degrees of anticipation that may be realized by such a decision. Last but not least, an exemplary class of dynamic decision problems from the field of vehicle routing is considered for detailed investigation.

Going all the way from understanding the nature of dynamic decision making to realization of anticipatory optimization at its best involves certain challenges. Facing these challenges has been my everyday experience throughout the past years of research. I am grateful for this experience as it comprises numerous fascinating insights into the subject of dynamic decision making and beyond.

Braunschweig, Germany, May 2011 *Stephan Meisel*

Preface

Acknowledgements

My research has been accompanied in different ways by a number of persons, whom I want to thank individually.

First of all, I would like to thank Professor Dirk Mattfeld for his support, for his valuable comments and for the liberty he granted me with respect to my research.

I would also like to thank both Professor Jürgen Branke and Professor Christian Bierwirth for the interest they showed in my work. Moreover, my thanks go to Uli Suppa, my former student, then colleague and friend, who has always been available for discussion and who contributed to my work in many ways. Furthermore I would like to thank Richard Verhoeven for his availability and for the many helpful comments he made.

Apart from all the fascinating scientific insights, an important lesson learned from the past years of research is that family and friendship are great gifts. Thus, I would like to express my particular gratefulness to my parents Renate and Karl-Heinz, to my sister Julia as well as to my many close friends, each of whom has a considerable share in this research.

Stephan Meisel

Contents

Chapter 1
Introduction

Economic operations are constantly subject to a large number of influences. Globalization, the progress of information technology as well as increasing resource prices are among the major influences of the past years. They raise both new challenges and opportunities for operations within a company. In addition, they open whole new contexts of economic activity. In particular these influences lead to operations under increasingly dynamic and uncertain conditions. The need for concepts and planning methods enabling the efficient conduction of such operations is a direct consequence. A company must respond to dynamic and uncertain circumstances by anticipation of future events in order to maintain efficiency.

Recent developments triggered by globalization, information technology and resource prices within a representative number of operational contexts are outlined in the following Sect. 1.1. Each of these contexts entails a number of decision problems resulting from the need for efficient operations subject to dynamics and uncertainty. In Sect. 1.2 an example decision problem is provided for each of the example contexts. The principle of anticipatory optimization and its relevance for solution of such decision problems is illustrated in Sect. 1.3. Section 1.4 gives an outline of the following chapters.

1.1 Recent Economic Developments

New developments can be observed in a broad variety of operational contexts. Transportation is one of the contexts strongly affected by each of the major influences afore mentioned. The importance of transportation increases rapidly due to decentralized supply chains in globalized markets. In addition, a higher volume and a higher frequency of transportation operations is triggered by both customer orders via internet and the growing significance of field services. Punctuality and responsiveness are becoming key features for satisfaction of delivery agreements and service guarantees as they appear in areas like just in time production or courier services.

The growing volume of transportation operations coincides with the rapid increase of the oil price. As a consequence operations are subject to heavy cost pressure. This pressure reinforces the use of technological innovations in order to gain efficiency. Hence, technologies like mobile communication, sensor and positioning devices as well as high performant data processing are gradually changing the mode of planning and implementation of transportation operations.

Similar tendencies are present in the context of manufacturing. New techologies enable the evolution from traditional mass production towards mass customization. Modern production processes must deal with an increasing number of customers issuing orders via internet. Supporting technologies like body scanners in clothing retail stores or web based CAD software for customers of furniture factories lead to a large number of product variants. Efficient mass production of customized products requires highly flexible and responsive production processes.

The need for flexibility in manufacturing is also raised by product diversification due to globalization. The presence of a company in a large number of domestic markets often requires many variants of a basic product. Production processes must cope with different types of country specific packaging or sizing for example. Flexibility and responsiveness become even more important in global companys relying on decentralized supply chains. In this case production processes must cope with an increased risk of delayed supplies in addition to a large number of variants.

However, new developments do not occur only in contexts dealing with material resources. The profound impact of technology and globalization on finance is a well-known example. Electronic trading systems and online banking allow for financial transactions at any time. In combination with reduced transaction costs due to liberalization and deregulation of capital markets highly dynamic acquisition and disposition of assets is becoming a common behavior of investors.

In addition relaxed investment restrictions and free cross-border capital flows facilitate international investments and allow for globally diversified portfolios. Thus, the number of investment options has been growing significantly and raises the question of how to adapt a portfolio to the dynamics of global markets. New techniques and methods for efficient asset management are both supported and enabled to a large degree by the increased capability of modern high performance computers.

A whole new set of operational planning problems has been induced in the context of energy management. Both the continuous rise of the oil price and the increasing environmental damage due to conventional energy production and due to the use of fossil fuels have been leading to a focus on renewable energy. A growing number of political directives on the use of sunlight, wind, rain and tides for energy generation appear. As an example the US Department of Energy aims at increasing the contribution of wind energy to US power supply up to twenty percent until 2030.

However, generating energy from natural resources tends to be both expensive and intermittent. As an example a company running a wind farm must develop plans for acting profitably within the energy market in spite of uncertain future wind conditions and high energy storage costs. Similarly, energy dispatching in electrical power grids is facing new challenges because of intermittent natural energy sources.

1.2 Examples of Emerging Problems

Transportation, manufacturing, finance and energy are only a subset of the operational contexts increasingly requiring repeated reaction to dynamics and uncertainty. In each of the contexts, reacting while maintaining efficiency finally boils down to decision problems that are hard to solve. These problems are important. The profit of a company may crucially depend on the ability to solve them. A number of illustrative decision problems emerging from different contexts are introduced in the following.

Less–than–Truckload Shipping

A vehicle represents a common type of resource to be managed efficiently in transportation. In the context of less–than–truckload (LTL) shipping a vehicle is used for collecting items from different customer locations. The items are consolidated on the vehicle. Finally the vehicle is heading for a hub where the items are transshipped for long distance transportation. As items in LTL shipping are both small and lightweight vehicle capacity limitations do not cause operational difficulties.

An LTL company aims at using a vehicle for serving as many customers as possible within the fixed period of time of a driver's shift. However, many customer requests may be unknown to the company at the beginning of the shift. Late requesting customers may call in while the vehicle already is en route. Each of the late requests must be either accepted or rejected for same day service by the company. An accepted request causes additional travel time for the vehicle. Each time additional requests are accepted the corresponding customer address must be communicated to the driver along with rerouting instructions.

Receiving new customer requests and communicating information updates to the driver can be done without problems due to modern information technology. However taking both acceptance decisions and rerouting decisions is often nontrivial with respect to the aim of maximization of the total amount of requests served within the current shift. On the one hand acceptance of a late request seems beneficial as it increases profits up to now. On the other hand additional travel time due to an acceptance decision possibly enforces the rejection of more than one request in the future.

Future requests are unknown but must be anticipated for making efficient decisions. As a basis for decisions a company may resort to knowledge about potential customer locations, travel time estimates as well as information about the current location of the vehicle possibly identified by GPS. Moreover the company may derive a request probability for each of the customers by analysis of customer records. However, the question of how to derive efficient decisions from this information remains.

Build to Order

In a manufacturing environment resources are often equivalent to machines or tools. Creation of a specified product requires a certain number of subsequent steps tied to a type of resource each. Often the order of production steps is strictly predefined. The types of resources required as well as the amount of time a resource is occupied for production heavily depend on the specification of the resulting product.

In many cases a customer is granted a deadline for receiving his product. Hence, manufacturers must be reliable, i.e., they must aim at the minimization of tardiness of production jobs. However, maintaining deadlines crucially depends on the scheduling decisions for production jobs. Such decisions must be made subject to a limited number of available resources as well as in many cases subject to the constraint of not canceling a job being processed.

Scheduling decisions resulting in minimal tardiness are hard to make particularly in the context of mass customization. The large number of possible product specifications implies a high variation of resource occupation times between different production jobs. In addition new jobs appear while others are being either processed or scheduled for production.

Releasing a job as early as possible is definitely a good decision with respect to its deadline. Nevertheless an early release may lead to inflexibilty and cause a massive amount of tardiness of jobs appearing in the future. As a consequence rescheduling decisions must be made if new production jobs appear. Information about type and probability of future jobs must be considered when taking scheduling decisions now in order to minimize costs due to tardiness.

Asset Allocation

The management of monetary resources raises financial planning problems. Asset allocation is a common type of financial planning problem. An asset allocation problem appears if an initial amount of money must be invested into different classes of financial assets. An investor compiles a portfolio of assets by deciding on the fraction of money to be invested into each possible class. The resulting portfolio must maximize the investor's wealth by the end of a predefined planning horizon. Additionally considering risk aversion leads to the objective of maximizing the sum of discounted returns over the horizon.

Obviously the terminal wealth depends on asset prices realized throughout the planning horizon. Unfortunately, prices typically are both non deterministic and varying over time. As a consequence rebalancing the portfolio by selling and buying assets may be benefical. The tendency towards rebalancing even increases in the presence of non prohibitive transaction costs. In particular, decisions to sell and buy assets are possible at any time throughout the planning horizon subject to asset specific transaction costs. Such costs of course represent the major drawback of a rebalancing decision.

Information acquisition may support beneficial decisions. An investor may derive probabilities representing the variation of single asset returns subject to correlations both across assets and across time. However, optimal rebalancing decisions will still be nontrivial especially in the case of a large number of asset classes. Moreover the tradeoff between decisions leading to a high amount of discounted wealth now and decisions taking into account transaction costs now for possible higher future returns must be resolved.

Wind Farm Management

Managing energy resources turns out to be challenging in the context of renewable energies. Wind farm operators face the problem of uncertain future wind conditions. An operator must periodically make an advance commitment of an amount of wind energy to the spot market. In case of generation of more energy than committed, the surplus is transferred to the operator's storage device. Stored energy is retrieved if wind conditions lead to underproduction with respect to the commitment made before. Both storage and retrieval operations are causing costs by reducing the amount of energy via conversion losses.

It is possible that the amount of energy on stock is not sufficient for balancing a false estimation of wind conditions. Then, an operator having committed too much to the spot market is subject to high penalty costs resulting from the need to buy energy from the balancing market. The balancing market is characterized by high energy prices and only comes into play if commitments cannot be satisfied.

The operating company is interested in maximizing the total sum of profits from energy sales over time. To this end a new amount of energy is committed to the spot market repeatedly. Before a commitment is made the operator may consult probability distributions about the energy generated in the following time period. Further, energy prices in the spot market and in the balancing market may be estimated before making a commitment decision. However, commitment decisions are interdependent and remain difficult to make. Committing less energy than produced in the upcoming period reduces profits due to conversion losses. Committing too much energy may lead to high penalties for drawing on the balancing market. As a byproduct each of the commitment decisions directly influences the amount of energy available for balancing underproduction in the next period.

Less–than–truckload shipping, build to order manufacturing, the allocation of assets and wind farm management seem to result into very diverse types of decision problems at first glance. Nevertheless, fundamental similarities exist. These similarities as well as their major implications are outlined in the next section.

1.3 Problem Similarities and Implications

The decision problems discussed represent a new type of problem within their particular operational context. From a decision theoretic point of view each of the problems falls into the category of dynamic decision problems. Accordingly a number of significant similarities exist.

Each of the problems is characterized by a sequence of decisions distributed over a given period of time. In any case a single decision determines the usage of one or more resources, e.g., a customer location is assigned to a vehicle, a job is assigned to a machine, a transaction is assigned to an amount of money or a commitment is assigned to an amount of energy.

A single decision is made based on the current knowledge about resources and about the environment. In each of the examples this knowledge is determined by both the precedent decision and by additional events occuring independently of the decision maker's influence. Any decision creates new knowledge by modification of resource characteristics, as e.g. position and availability of a vehicle, machine occupancies, money disposal or utilization of energy. Future decisions are subject to the results of such a modification as well as subject to the results of changing conditions due to, e.g. late customer requests, new production jobs or variations of asset prices and wind intensity.

In each of the examples the sequence of interdependent decisions is carried out for achievement of a superordinate goal, i.e. for maximization of the total number of customers served, for minimization of the overall tardiness of jobs, for maximization of the sum of discounted returns and for the maximization of profits from energy sales. A single decision generates a distinct contribution to this goal. At the same time it influences the conditions of the contributions of the following decisions. The sum of the single contributions of a sequence represents its total value.

The fact that the problems can be identified as dynamic decision problems does not answer the question of how to actually solve them, i.e., the question of how to realize dynamic decision making. Of course dynamic decision making must be realized as good as possible with respect to the superordinate goal. Single decisions should be optimal. However, optimality of a single decision requires consideration of both its immediate contribution and its impact on successive decisions. A relatively large number of decisions must be coordinated throughout the time horizon. The search for an optimal decision must anticipate a decision's future consequences.

This emerging class of operational problems in an economic context shares particular characteristics. These characteristics require an improved and extended approach to dynamic decision making. One of the major challenges originates from the fact that even without considering anticipation of future consequences, making a single decision involves a decision problem hard to solve. Traditionally such single problems have been treated within the field of operations research.

The economic relevance of the problems raises the need for good solutions. But what does it take to actually get to a good solution? This question is addressed in the following. Within the subsequent chapters, the building blocks of anticipatory optimization are deduced and integrated successively. In addition the practical impact of dynamic decision making by anticipatory optimization is exemplified on the basis of example problems as introduced in the previous section. A guideline of the following chapters is provided in the next section.

1.4 Outline of the Following Chapters

From a high-level perspective the remaining chapters fall into three major steps. At first, Chaps. 2 to 5 develop both the conceptual foundation and the methodological basis of anticipatory optimization for dynamic decision making. Subsequently Chap. 6 provides a systematic exposure of the field of dynamic vehicle routing, which represents an exemplary class of dynamic decision problems. The third major step consists in transferring the generic concept of anticipatory optimization to application. Chapters 7 and 8 focus on anticipatory optimization for a family of dynamic vehicle routing problems, while Chap. 9 considers application from a more general perspective. Eventually, concluding remarks with respect to each of the three steps are provided in Chap. 10. The single contributions of Chaps. 2 to 9 are as follows:

Chapter 2 defines anticipatory optimization for dynamic decision making. It introduces the concepts of "dynamic decision making", "optimization" as well as "anticipation" and unites them from a conceptual point of view. The chapter provides a classification of anticipatory decisions according to their respective degree of anticipation. Moreover, it identifies the different types of optimization problems that result from different degrees of anticipation. Throughout the remaining chapters, the degree of anticipation is considered as the primary characteristic of an approach to anticipatory optimization for dynamic decision making.

Chapter 3 covers a broad family of methods for realization of perfect anticipatory decisions. Starting from the classic approaches of dynamic programming, the chapter gradually develops the many flavors of simulation-based methods for perfect anticipation. These methods are subsumed under the actor-critic framework, which provides an important point of reference for the remaining chapters. Finally, Chap. 3 raises the need for a tradeoff between degree of anticipation and computational efficiency by discussing the limited effectiveness of perfect anticipation.

Chapter 4 introduces the concept of "data mining" and integrates optimization and data mining into one unified methodology. The latter provides the conceptual foundation of a generalized actor-critic framework for realization of lower degrees of anticipation. Beyond a description of the elementary building blocks of this foundation, Chap. 4 identifies the synergies of optimization and data mining on a general level and illustrates these synergies by examples.

Chapter 5 integrates the methodology of Chap. 4 and the actor-critic framework of Chap. 3. The resulting family of procedures allows for approximation of perfect anticipatory decisions at a feasible computational cost. Approximate anticipation still takes into account the nature of dynamic decision problems by relying on the principles introduced in Chap. 3. However, at the same time it permits various levels of approximation via the unified methodology of Chap. 4.

Chapter 6 presents the field of dynamic vehicle routing as an exemplary class of dynamic decision problems. It motivates the need for anticipatory decisions in vehicle routing and provides an overview of state-of-the-art approaches to realization of such decisions. Although these approaches treat a variety of dynamic vehicle routing problems, they are consistently categorized according to their respective degree of anticipation. Moreover, the tremendous complexity of anticipatory vehicle routing is illustrated by means of a detailed analysis of the problem of dynamic routing of a service vehicle. This analysis also serves as a motivation of the following Chaps. 7 and 8.

Chapter 7 proposes different approaches to anticipatory optimization for dynamic routing of a service vehicle. These approaches are characterized by different degrees of anticipation and provide the basis of the computational study in Chap. 8. The particular focus of Chap. 7 is twofold. On the one hand, it analyzes perfect anticipation for dynamic routing of a service vehicle. On the other hand, Chap. 7 draws on the general methods of Chap. 5 and develops a number of procedures realizing approximate anticipation for dynamic routing of a service vehicle.

Chapter 8 provides an experimental comparison of the anticipatory procedures introduced in Chap. 7. To this end, a variety of instances of the problem of dynamic routing of a service vehicle is considered.

Chapter 9 discusses anticipatory optimization from a managerial point of view. The requirements implied by different degrees of anticipation are illustrated and criteria for selecting a certain degree of anticipation are developed.

Chapter 2
Basic Concepts and Definitions

This chapter defines the concepts representing the fundamental point of reference of anticipatory optimization for dynamic decision making. Section 2.1 summarizes the elements of dynamic decision making. In Sect. 2.2 optimization is defined as the procedure of finding a solution of a single decision problem. Section 2.3 introduces the concept of anticipation and exposes the scope and range of anticipatory optimization.

2.1 Dynamic Decision Making

Dynamic decision making is a well established part of both decision theory (e.g., French, 1988) and the field of decision analysis (e.g., Clemen and Reilly, 2001).[1] In line with these fields a basic dynamic decision making process is introduced in the following. Subsequently this process is embedded into the context of Markov decision processes.

2.1.1 A Basic Dynamic Decision Process

Dynamic decision making refers to a process consisting in a series of decisions in a temporal order. Figure 2.1 illustrates a dynamic decision process. At each decision time t a single decision d_t is determined based on the current knowledge about resources and other relevant attributes of a particular system under consideration. This knowledge is denoted as the state s_t of the system at time t.[2] A state s_t defines the set D_t of feasible decisions at time t.

[1] In the literature dynamic decision making is often subsumed under the more general categories of "multi-stage decision making", "sequential decision making" or "decision trees".

[2] More details about the concept of state as well as other concepts introduced in the following are provided in the next section.

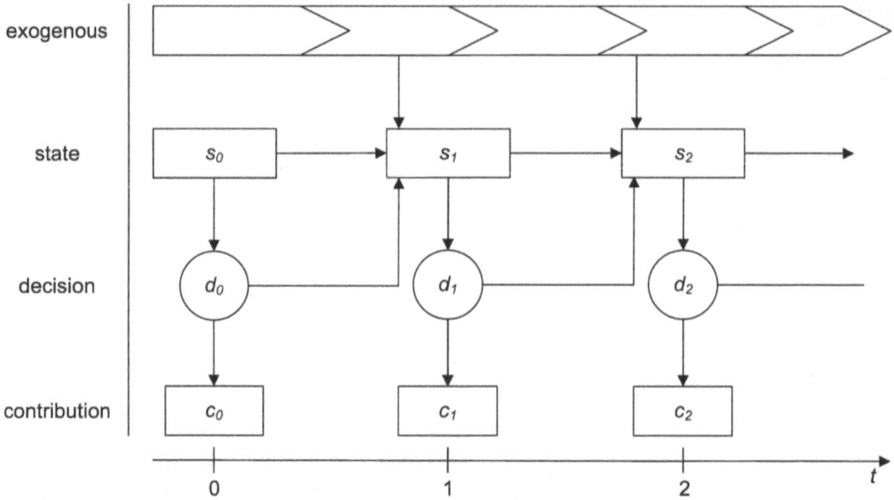

Fig. 2.1 A dynamic decision process.

A successor state $s_{t'}$, with $t' > t$, is subject to three potential determinants. In addition to s_t and d_t an exogenous process may affect $s_{t'}$. This process is perceived as a random influence by the decision maker. In general such a random influence is continuous in time. While decisions may basically be exerted continuously too, most dynamic decision processes coming from an economic background require taking action at selected points in time only. In the dynamic decision process of Fig. 2.1 decision times are numbered serially. Moreover in this process each decision results in a contribution c_t to one single overall objective of the decision maker. The contribution may be determined by d_t only. Additionally the state s_t and the exogenous process may each influence the magnitude of c_t.

The interaction of a decision d_t and the exogenous influence may be of different nature with respect to a state transition. For some problems the two influences can hardly be kept apart from each other. As an example think of a decision to act on a resource with the decision's impact on the resource's attributes being subject to uncertainty.[3] However, for many dynamic decision problems the influences of d_t and the exogenous process are not intertwined but can be considered separately.

Two important cases of a state transition are illustrated in Fig. 2.2. The example system considered in the figure comprises a set $S = \{1,\ldots,7\}$ of seven states. Each of the states can theoretically be reached at any decision time. The left hand side of Fig. 2.2 shows the special case of a state transition in the absence of an exogenous influence. As a consequence the transition becomes deterministic and the successor

[3] In such a case, some authors (e.g., Van Hentenryck and Bent, 2006) consider the uncertainty to result from an endogenous influence being inherent in a decision instead of coming from an exogenous source. However, even in the case the random influence is intertwined with the decision, it still remains an uncontrollable influence for the decision maker. For this reason we still consider such a random influence as exogenous.

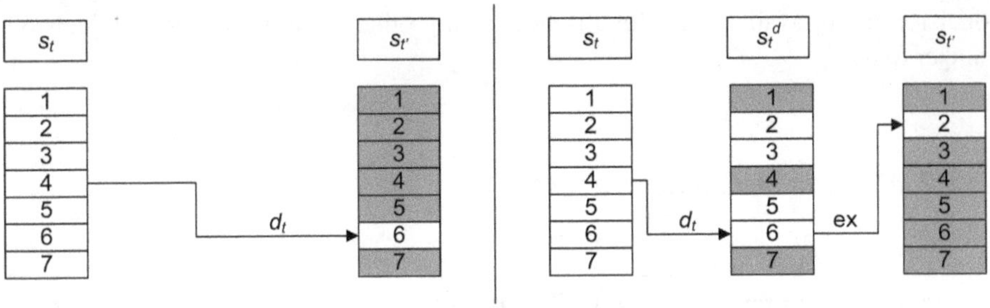

Fig. 2.2 State transition without exogenous influence (*left hand side*) and with an exogenous influence that is separable from the influence of a decision (*right hand side*). Unreachable states are *gray-shaded.*

state $s_{t'} = 6$ is a unique result of decision d_t taken in state $s_t = 4$. On the right hand side of the figure an exogenous process exists but its impact on the state transition can be separated from the impact of d_t. The latter is deterministic and becomes fully known as soon as the decision is made. Hence, starting from $s_t = 4$, decision d_t again leads to state six. Instantaneously the system shifts to a state $s_t^d = 6$ which is not necessarily equivalent to the system state $s_{t'}$ at the next decision time t'. The decision reduces the set of reachable states in t'. Yet $s_{t'}$ results from the random influence of the exogenous process (denoted as "ex"). The influence completes the state transition by leading from $s_t^d = 6$ to $s_{t'} = 2$. Note that $s_t^d = s_{t'}$ in the absence of an exogenous process.

Procedure 1: A generic outline of dynamic decision making

```
t ← 0
s0 ← initial state
repeat
    if (condition for next decision) then
        d_t ← DetermineNextDecision(s_t)
        initialize execution of d_t
    end
    t ← t+1
    s_t ← current state
until (t = 𝒯)
```

Procedure 1 provides a generic outline of dynamic decision making. Starting from an initial state s_0 at time $t = 0$ the procedure steps forward in time until a predefined end of the time horizon is reached. In addition to the length $\mathscr{T}$ of the time horizon, the condition for making the next decision is an important parameter of dynamic decision making. A new decision may be triggered by the occurence of a predefined set of events. Alternatively a new decision may be due continuously at any time t. This is the case in particular if "do nothing" is considered as a valid decision being available in the absence of any other event requiring the decision maker to act.

2.1.2 Markov Decision Processes

The characteristics of a state are one of the key features of the decision process described in the previous section. The current state s_t serves as the sole basis for making a decision d_t. A state transition and a contribution c_t both never depend on the predecessors of s_t. This independence of history is the main property characterizing the decision process as a Markov decision process (MDP). MDPs were originally derived from the theory of Markov chains by Bellman (1954, 1957b). In an MDP the state variable is the minimally dimensioned function of history that is both necessary and sufficient for identification of d_t, calculation of c_t as well as the state transition (Powell, 2007, Chap. 5).

Of course alternative models of dynamic decision making can be constructed. A different way of modeling the state of a system may for example result in decisions made under consideration of several past states. Moreover there may be dynamic decision processes that deviate from the process of the previous section with respect to the contributions c_t.[4] However, the theory of MDPs is sufficiently broad to allow for modeling most realistic dynamic decision making situations (Puterman, 2005, Sect. 1.1). In the following a dynamic decision process is considered as a Markov decision process $\{T, S, D, p, c\}$ with the subsequent properties:

- T denotes the set of possible decision times.[5] The length $\mathscr{T}$ of the time horizon considered for dynamic decision making determines the number of elements of T, i.e., $|T| = \mathscr{T}$. A decision may be made at each time $t \in T$. Alternatively some elements of T may not necessarily require a decision. Actual decision times may then be determined either randomly by the occurence of a certain event or deliberately by the decision maker. These decision times are contained in an ordered set $\tau = (\tau_0, \tau_1, \ldots, \tau_{|\tau|-1}) \subseteq T$ with τ_i making reference to the i-th element of τ.
- S denotes the set of all the states that may appear throughout the time horizon. As one state may appear at several decision times, the decision time is included into the state variable. Taking the decision time away from the state variable leads to a number $\mathscr{T}$ of sets S_t with $S_t \times \{t\} \subseteq S$. A set S_t comprises the states that can possibly appear at a specific decision time t.

[4] On the one hand a separate assignment of a contribution to a state transition may be hard to define in some cases. On the other hand an additive combination of the single contributions suitable for measuring the decision maker's overall success may be difficult to derive. The latter would be particularly challenging if a decision maker pursued a number of distinct objectives which might on top of it be competing. However, the more general field of multi-objective dynamic decision making will not be treated in the following.

[5] Decision times are often referred to as "decision epochs" in the literature.

- D denotes the set of decisions that may be exerted throughout the time horizon. A set $D_t(s_t) \subseteq D$ comprises the decisions available at time t with the current state being s_t.
- p denotes a function $p : S \times D \times S \mapsto [0 \dots 1]$ representing the impact of the random influence on a state transition. A value $p(s'|d,s)$ denotes the probability of a transition to s' if decision d is made in state s. Moreover, for a given point in time t, $p_t(s_{t'}|d_t, s_t)$ denotes the probability of a transition to state $s_{t'}$ at time t' if decision d_t is made in state s_t.
- c denotes a function representing the contribution emerging in the context of a state transition. The contribution may be either deterministic or random. In the deterministic case c may depend on both the current state s and the decision d, i.e. $c : S \times D \mapsto \mathbb{R}$. Alternatively c may exclusively be subject to d and s respectively. A random contribution occurs if c is affected by the exogenous process. In this case the actual contribution $\hat{c}(s,d,s')$, with s' being a state at t', becomes known not until the next decision time t'. Hence, a decision could be made based on a contribution function $c : S \times D \mapsto \mathbb{R}$ with $c(s,d) = \mathbb{E}[\hat{c}|s]$. Moreover, for each point in time t the function $c_t : (s_t, d_t)$ returns the contribution generated by making a decision d_t while being in state s_t.

A thorough overview of the different flavors of MDPs resulting primarily from a variation of the characteristics of T, S or D can be found in the book of Puterman (2005). A notable extension of the fundamental MDPs is given by the concept of partially observable Markov decision processes (POMDPs).[6] A POMDP represents a dynamic decision process if part of the state cannot be determined exactly. A general survey of POMDP applications comprising some dynamic decision problems is provided by Cassandra (1998).

2.2 Optimization

Although optimization is used in various branches of applied mathematics and statistics, it is particularly associated with the field of operations research. (Beale, 1988, Sect. 1.1) In the following, optimization is introduced from this perspective. In Sect. 2.2.1 the connection between dynamic decision making and optimization is established. Then the basic types of optimization techniques for solution of such optimization problems are discussed in Sect. 2.2.2.

2.2.1 Optimization Problems

Optimization problems are introduced in slightly different styles in the literature. Nemhauser et al (1989) state that optimization deals with problems of minimizing

---

[6] POMDPs were originally introduced by Astrom (1965) and Aoki (1965).

or maximizing a function of several variables usually subject to equality and/or inequality constraints. The connection between optimization and decision making is pointed out explicitly by Chong and Zak (2008), introducing optimization as central to any problem involving decision making. Therefore an optimization problem is identified with the task of selecting the best out of a number of alternative decisions in the sense of a given objective function. The presence of an objective function representing a preference relation among alternative decisions makes the coincidence of a decision problem and an optimization problem.

An optimization problem $P = (X, f)$ consists of a set X of feasible solutions (decisions) as well as a map $f : X \mapsto \mathbb{R}$.[7] The map f is often referred to as objective funtion, cost function or evaluation function. In the following f is denoted as the objective function and an optimization problem is formulated as a maximization problem.[8] Accordingly the aim of optimization is determination of an optimal solution $x^\star \in X$, with

$$\forall x \in X : f(x^\star) \geq f(x) . \tag{2.1}$$

Elements $x^\star \in X$ for which Condition 2.1 is satisfied are referred to as global optima. As identification of a global optimum may be very difficult with respect to a given optimization problem, the aim of optimization is often reformulated in terms of local optima. A local optimum must not necessarily be identical to a global optimum. Nevertheless, it still represents a solution that is of relatively high quality. More precisely, a local optimum x^* is defined with respect to a set of adjacent solutions, referred to as a neighborhood $N(x^*) \subseteq X$ of x^*.[9] The quality of a local optimum x^* may be worse than the quality of other $x \in X$. Nevertheless x^* must satisfy

$$\forall x \in N(x^*) : f(x^*) \geq f(x) . \tag{2.2}$$

In the context of a Markov decision process an optimization problem may be associated with the decision problem at a single point in time t. Let D_t be the set of possible decisions at time t and assume c_t not to depend on a random influence. If a decision maker exclusively cares about the current situation at t, his aim is maximization of c_t. In other words, he faces the deterministic optimization problem $P_t = (D_t(s_t),\ c_t(s_t, \cdot))$. P_t becomes a stochastic optimization problem if c_t is subject to the random influence of an exogenous process, i.e. $P_t = \Big(D_t(s_t),\ \mathbb{E}\big[\hat{c}_t(s_t, \cdot, s_{t'})|s_t\big]\Big)$.

If the decision maker's goal was maximization of every single contribution regardless of the remaining decision process, dynamic decision making would be identical to solving a sequence of optimization problems of type P_t. The same would be true if the goal was maximization of the sum of contributions but single decisions were independent of each other. Moreover, decision processes with interdependent

---

[7] In the presence of multiple distinct objectives the range is extended to $\mathbb{R}^n$. However this presentation focuses on single objective optimization. Thus the range of f is assumed to be $\mathbb{R}$.

[8] This convention does not imply a restriction because an equivalent maximization problem can be formulated for any given minimization problem.

[9] Note the notational difference between a global optimum $x^\star$ and a local optimum x^*.

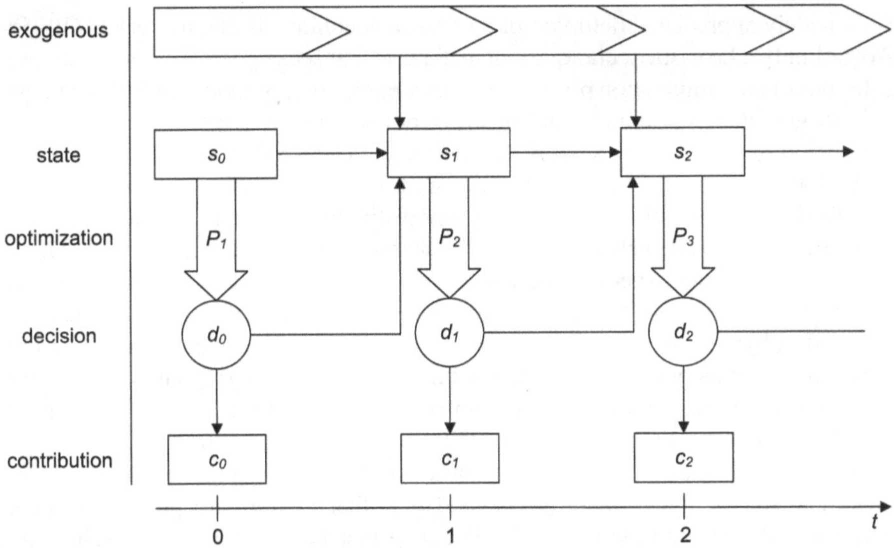

Fig. 2.3 Optimization in the context of dynamic decision making.

decisions may exist, for which the distinct goals of maximization of each single contribution and maximization of the sum of contributions coincide. However this is not the case in general. The question of which objective function to choose for an optimization problem P_t occuring in dynamic decision making is addressed in Sect. 2.3.1.

Figure 2.3 illustrates the occurence of optimization problems P_t within dynamic decision making. Provided an objective function is formulated, optimization is conducted at each decision time t by means of optimization techniques. The basic categories of optimization techniques are discussed in the next section.

2.2.2 *Optimization Techniques*

Two fundamental alternatives of algorithmic handling of an optimization problem exist. On the one hand many problems can be solved by use of an exact algorithm. An exact algorithm definitely leads to a global optimum. On the other hand, heuristic techniques may be applied for finding a solution. In contrast to an exact algorithm a heuristic technique cannot guarantee for a global optimum in the majority of cases. As a consequence, strictly speaking the term "optimization" is out of place with respect to heuristics. However, the use of both exact and heuristic approaches is considered as optimization in the literature. Some authors (e.g., Aarts and Korst, 1989) even explicitly stress the fact that there is a smooth transition between the notion of an exact algorithm and the notion of a heuristic algorithm.

A widely appreciated defintion of a heuristic technique is due to Reeves (1995). Accordingly a heuristic technique is a method which seeks good (i.e. near-optimal) solutions of an optimization problem at a reasonable computation cost without being able to guarantee optimality, and possibly not feasibility. Further, it may not be possible to state how close a heuristic solution is to a global optimum.

Various categories of heuristic techniques are distinguished in the literature (see, e.g., Burke and Kendall, 2005, Chap. 1). The distinction between a problem-specific heuristic and a metaheuristic is of particular importance. The quality of a solution derived by a problem-specific heuristic depends to a very large amount on a priori available insights about the decision problem under consideration. These insights are used as hypotheses for tailoring an algorithm which often cannot be applied to decision problems other than the one at hand. Note that tailoring an algorithm for an optimization problem $P = (X, f)$ can often be understood as defining a proxy $P' = (\tilde{X}, \tilde{f})$ whose solution is assumed to be good with respect to P.

The term metaheuristic refers to a master strategy that guides and modifies another heuristic to produce solutions beyond those that are normally generated by the heuristic (Glover and Laguna, 1997). Hence, a metaheuristic does not exclusively rely on hypotheses deduced from the decision problem at hand. Instead a generic hypothesis is additionally introduced promising to be a good way of guiding the underlying problem-specific heuristic. Most metaheuristics are characterized by a number of parameters to be tuned to the decision problem at hand. Well-known examples of metaheuristcs are simulated annealing (Kirkpatrick et al, 1983), tabu search (Glover and Laguna, 1997), genetic algorithms (Goldberg, 1989) and the greedy randomized adaptive search procedures (Feo and Resende, 1995).

A heuristic algorithm can be constructed for almost any optimization problem. The prevalent application field are problems for which an exact algorithm offering polynomial runtime is not available. Moreover heuristic algorithms are often preferred to exact ones due to higher flexibility with respect to runtime.

2.3 Anticipation

The notion of "anticipation" has been a subject of scientific discussion within recent decades. Much attention was given to Rosen (1985) defining an anticipatory system as a system containing a predictive model of itself and/or of its environment, which allows it to change state at an instant in accord with the model's predictions pertaining to a later instant. Butz et al (2003) define anticipatory behavior as a process, or behavior, that does not only depend on past and present but also on predictions, expectations, or beliefs about the future. From a literal point of view the meaning of the term "to anticipate something" may be defined as "to expect this thing to happen and be ready for it" (Longman, 2003) or similarly as "to forsee and deal with this thing in advance" (Merriam-Webster, 2003).

In dynamic decision making the remaining decision process is to be anticipated at each decision time t. A decision at time t provides the means of dealing with the remaining decision process. In Sect. 2.3.1 a criterion for making an anticipatory decision within a dynamic decision process is derived. Subsequently different degrees of actually meeting this criterion are discussed in Sect. 2.3.2.

2.3.1 Anticipatory Decisions

A single decision within a dynamic decision process has two important roles with respect to the decision maker's overall objective. On the one hand d_t causes an immediate contribution c_t. On the other hand, d_t has a significant influence on the successor state $s_{t'}$. That latter represents a major determinant of the remaining decision process and with it of the future contributions to the overall objective. An anticipatory decision aims at an optimal tradeoff between both the immediate contribution and the impact on future contributions. This claim for optimality must be reflected in the formulation of an optimization problem P_t. Such a formulation leading to an optimal anticipatory decision at a decision time t is derived in the following.

A decision maker aims at maximization of the sum C of contributions resulting from a sequence $(d_0, d_1, \ldots, d_{\mathcal{T}-1}) \in D^{|T|}$ of single decisions, i.e.

$$C = \max_{D^{|T|}} \sum_{t \in T} c_t(s_t, d_t) \, . \tag{2.3}$$

However, taking into account the random influence of an exogenous process, the overall objective of a decision maker must be formulated as maximization of the expected sum of contributions

$$\bar{C} = \max_{D^{|\tau|}} \mathbb{E}\big[\sum_{t \in \tau} c_t(s_t, d_t) | s_0\big] \, . \tag{2.4}$$

For $|T| = 1$ the decision process comprises a single optmization problem $P_0 = (D_0, c_0)$. In this case, $\bar{C}$ may be considered as the value $V_0(s_0)$ of the initial state s_0 for the decision maker. As a consequence

$$V_0(s_0) = \bar{C} = \max_{d_0 \in D_0(s_0)} c_0(s_0, d_0) \, . \tag{2.5}$$

The concept of the value $V_t(s_t)$ of a state s as the expected sum $\bar{C}$ of contributions achieved subject to optimal decisions in the remaining decision process can be extended to arbitrary MDPs with $|\tau| \geq 1$. Representing the expectation in terms of the state transition probabilities $p_t(s_{t'}|s_t, d_t)$ and assuming $\forall t : S_t = \{1, 2, \ldots, k_t\}$, the value V_t of a state s_t subject to optimal decisions within the remaining process is

$$V_t(s_t) = \max_{d_t \in D_t(s_t)} \left[c_t(s_t, d_t) + \sum_{s_{t'}=1}^{k_{t'}} p_t(s_{t'}|s_t, d_t) V_{t'}(s_{t'}) \right] , \tag{2.6}$$

with $\forall s_{\mathcal{T}} \in S_{\mathcal{T}} : V_{\mathcal{T}}(s_{\mathcal{T}}) = 0$. Equation 2.6 defines the value of a state recursively. The set of equations defining the values of all the states $s_t \in S_t$ for each t is known as Bellman's equations (Bellman, 1957a).[10] They provide the objective function for an optimal anticipatory decision

$$d_t^\star = \arg \max_{d_t \in D_t(s_t)} \left[c_t(s_t, d_t) + \sum_{s_{t'}=1}^{k_{t'}} p_t(s_{t'}, s_t, d_t) V_{t'}(s_{t'}) \right] . \tag{2.7}$$

Such a decision $d_t^\star$ realizes the optimal tradeoff between the immediate contribution c_t and, via the values $V_{t'}$ of the successor states, the impact on future contributions. The corresponding optimization problem P_t to solve at time t in state s_t is

$$P_t = \Big(D_t(s_t),\; c_t(s_t, d_t) + \mathbb{E}\left[V_{t'}(s_{t'}) | s_t \right] \Big) . \tag{2.8}$$

Note that the distribution of $s_{t'}$ depends on d_t. P_t represents an ordinary optimization problem as introduced in Sect. 2.2.1. Basically the quality of a solution of P_t is determined by the type of optimization technique applied (see Sect. 2.2.2). However, in the next section a more fine-grained classification of solution quality is established allowing for a distinguished characterization of anticipatory decisions.

2.3.2 Degrees of Anticipation

A detailed classification of anticipatory decisions exists within the broad framework of distributed decision making (Schneeweiss, 2003). From this point of view a dynamic decision process is considered as a decision time hierarchy, which is a special case of so called organizatorial distributed decision making.[11] Anticipatory decisions are categorized according to the prototypes of anticipation summarized in Fig. 2.4.[12]

In the following these prototypes are illustrated by means of a dynamic decision process with $\mathcal{T} = 2$ and $\tau = \{0, 1\}$.[13] This decision process comprises two decision problems P_0 and P_1, which can both be stated according to Equation 2.8. Anticipation is relevant only within P_0 as the value $V_2(s_2)$ of a terminal state is both deterministic and equal to 0. Solution of P_0 requires the value function $V_1(s_1)$. At $t = 1$ the value of a state $s_1 \in S_1$ is identical to the contribution received by making an optimal decision $d_1^\star$, i.e., $V_1(s_1) = c_1(s_1, d_1^\star)$. Equation 2.9 shows the resulting formulation of P_0. Note that the value $V_1(s_1)$ of a state s_1 is calculated subject to d_0

[10] The formulation of Bellman's equations depends on the properties of the underlying Markov decision process. Equation 2.6 is based on the assumption of countable sets T and S as well as on the condition $\forall s_{\mathcal{T}} \in S_{\mathcal{T}} : V_{\mathcal{T}}(s_{\mathcal{T}}) = 0$. Formulations of Bellman's equations for MDPs with different properties can be found in Puterman (2005).

[11] Within the framework of distributed decision making, a major distinction is made between "constructional" settings and "organizatorial" settings. The former are characterized by different decisions to be made both simultaneously and under consideration of the same pieces of information. The latter imply information asymmetry with respect to different decisions.

[12] Fig. 2.4 is in the style of Fig. 2.6 in Schneeweiss (2003).

[13] These assumptions are made for illustrative purposes only.

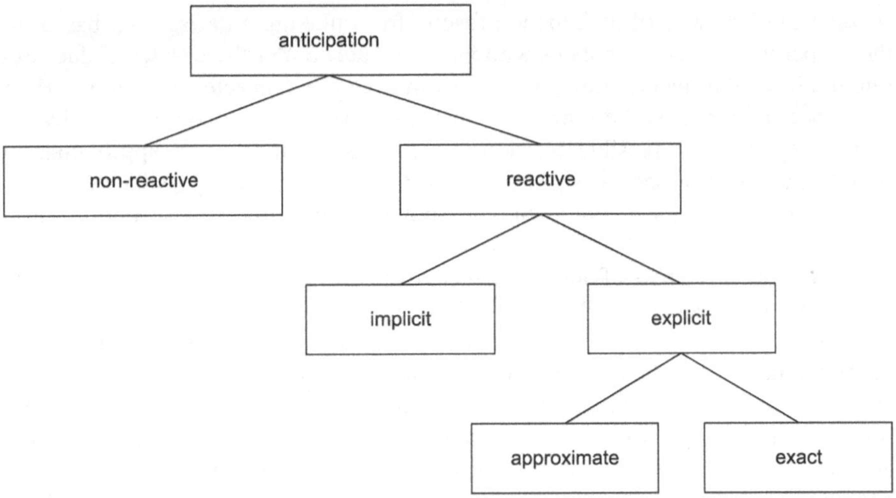

Fig. 2.4 Prototypes of anticipation.

as $s_1 = s_1(d_0)$ and $d_1^\star = d_1^\star(s_1)$. Solving P_0 by use of an exact algorithm leads to a *perfect*[14] anticipatory decision $d_0^\star$. Perfect anticipation requires the state values to be available *exactly*. However, exact determination of these values is hard to realize in many cases. Both computational limitations (see Sect. 3.4) and a lack of information needed for characterization of the MDP may occur. For this reason a tradeoff between the quality of anticipation and the feasibility of the optimization approach is required.

One realization of this tradeoff results from substitution of P_0 for a proxy P_0' as given by Equation 2.10. P_0' still considers the remaining decision process explicitly by taking account of the values of successor states in the objective function. Yet, it requires only *approximate* state values within the objective function. An approximate state value may be derived in various ways. One straightforward possibility comprises substituting $d_1^\star$ for a local optimum d_1^*, i.e., $\tilde{V}_1(s_1) = c_1(s_1, d_1^*)$.

$$P_0 = \Big(D_0(s_0),\ c_0(s_0,d_0) + \mathbb{E}\big[V_1(s_1)|s_0\big]\Big) \tag{2.9}$$

$$P_0' = \Big(D_0(s_0),\ c_0(s_0,d_0) + \mathbb{E}\big[\tilde{V}_1(s_1)|s_0\big]\Big) \tag{2.10}$$

$$P_0'' = \Big(\tilde{D}_0(s_0,\tilde{V}_1),\ c_0(s_0,d_0)\Big) \tag{2.11}$$

$$P_0''' = \Big(\tilde{D}_0(s_0),\ f_0(d_0)\Big) \tag{2.12}$$

[14] The term perfect may be misinterpreted in this context. A decision is referred to as perfect with respect to the objective as stated in Equation 2.4. Though being "perfect" such a decision will only rarely result in the ideal of a maximization of the sum of contributions as stated in Equation 2.3.

A more *implicit* way of anticipation results from making a decision d_0 based on the properties of successor states without taking account of these states' values explicitly. Instead of incorporating the state values into the objective function of P_0, a deliberate influence on the remaining decision process is exerted implicitly by restriction of the set of feasible decisions D_0. A possibly quite coarse approximation of the value function serves as a basis for excluding a number of possible decisions, i.e., $\tilde{D}_0(s_0) = \tilde{D}_0(s_0, \tilde{V}_1)$. Equation 2.11 shows the resulting type of optimization problem.

Though their degrees of anticipation are different, anticipatory decisions resulting from problems of type P_0 through P_0'' share the common feature of reacting to the remaining decision process. The reaction consists in adjusting a decision to the current situation at time t under explicit or implicit consideration of the values of succesor states. As an alternative, an anticipatory decision d_0 may account for the remaining process on a more general *non-reactive* level. A decision may be made according to a heuristic procedure incorporating general hypotheses about the decision process. For example a hypothesis may simply express the fact that something will occur in the future. Incorporating such hypotheses into an optimization problem P_0''' may result in both an approximation $\tilde{D}_0$ of the of set feasible decisions and a general objective function $f_0(d_0)$ (cf. Equation 2.12).

The leaves of the tree shown in Fig. 2.4 represent the prototypes of anticipation. The degree of anticipation associated with these fundamental categories increases from non-reactive approaches to perfect anticipation. The latter is characterized as being reactive as well as both explicit and exact. A variety of approaches to perfect anticipation is illustrated in the next chapter.

Chapter 3
Perfect Anticipation

A perfect anticipatory decision in the context of dynamic decision making may be achieved by solution of an optimization problem as stated in Eq. 2.8. However, formulation of such an optmization problem requires the availability of the values $V_{t'}(s_{t'})$ of successor states $s_{t'}$. In particular, a value function $V_t(s_t)$ must be known for each of the decision times t. Determination of these value functions corresponds to the solution of Bellman's equations as formulated as Eqs. 2.6.

The following sections comprise three categories of approaches to solving Bellman's equations. Each category assumes the presence of a single terminal state $s_{\mathcal{T}}$ with $V_{\mathcal{T}}(s_{\mathcal{T}}) = 0$.[1] Moreover each category is based on the assumption that for each possible initial state s_0 a sequence of decisions leading to $s_{\mathcal{T}}$ exists.

The first category of approaches is given by the elementary methods of dynamic programming (Sect. 3.1). These methods provide the foundation of the simulation based approaches known as forward dynamic programming. A number of important variations of forward dynamic programming are described in Sect. 3.2. Section 3.3 treats a category of simulation based approaches to perfect anticipation that can be applied if the transition function p cannot be specified explicitly. Each of the methods presented in this chapter enables perfect anticipation by solution of Bellman's equations. However, for most dynamic decision problems of practical relevance these methods imply enormous computational requirements. Thus, Sect. 3.4 points out the sources of the computational burden needed for perfect anticipation.

3.1 Dynamic Programming

This section introduces the elementary methods of dynamic programming. The iterative methods of Sects. 3.1.1 and 3.1.2 are generalized in Sect. 3.1.3 and contrasted with the mathematical programming approach (Sect. 3.1.4). The presentation

[1] In case more than one terminal state exists, one additional state may be introduced as the deterministic successor of each terminal state.

assumes time to be included into the state variable. This assumption reduces the notational complexity and makes the basic idea behind the methods come forward more clearly.

3.1.1 Value Iteration

Value iteration is known as the earliest method of dynamic programming. The procedure was originally introduced by Bellman (1957a). It derives the value $V(s)$ of each state $s \in S$ iteratively. Several algorithmic variations exist. The most intuitive among them implements the concept of backward induction. In an iteration n, each state s is assigned a value $V^n(s)$. $V^n(s)$ is the maximum expected sum of contributions that can be realized by taking at most n subsequent decisions for getting from s to the terminal state $s_{\mathscr{T}}$. In case $s_{\mathscr{T}}$ cannot be reached starting from s by making n decisions, the value of s is left unchanged in iteration n, i.e., $V^n(s) = V^{n-1}(s)$. The value of the terminal state is set to $V^0_{\mathscr{T}}(s_{\mathscr{T}}) = 0$ while the values of the other states $s \in S$ are initialized as $V^0(s) = -\infty$. Assuming $S = \{1,2,\ldots,k\}$ the value $V^n(s)$ of s in iteration n is updated according to

$$V^n(s) = \max_{d \in D(s)} \left[c(s,d) + \sum_{s'=1}^{k} p(s,s',d)\, V^{n-1}(s')\right] . \tag{3.1}$$

The procedure returns the true values $V(s) = V^{\mathscr{T}-1}(s)$ after $\mathscr{T} - 1$ iterations. Procedure 2 represents a variation of the original value iteration procedure. It can be shown (Bertsekas, 2007) that this variation letting $V^0(s) = 0$ also converges to the true values $V(s)$.[2] However, no matter how the values are initialized, convergence of value iteration often implies a high computational burden. Note that typically the number of states is very large in dynamic decision problems and that each update of a value requires exact solution of a stochastic optimization problem.

Procedure 2: Value Iteration

$n \leftarrow 0$
$\forall s \in S : V^0(s) \leftarrow 0$
repeat
  $n{+}{+}$
  for $(s \leftarrow 1, s \leq k, s{+}{+})$ **do**
    $V^n(s) \leftarrow \max_{d \in D(s)} \left[c(s,d) + \mathbb{E}\left[V^{n-1}(s') | s\right]\right]$
  end
until $(\forall s \in S : V^n(s) = V^{n-1}(s))$

Early termination can be provoked by introduction of a threshold ε. With it value iteration terminates if for each state the change of the (estimated) value is less than

[2] In particular Bertsekas (2007) proves convergence subject to an arbitrary initialization of $V(s)$.

ε between two subsequent iterations, i.e., the criterion for termination becomes $\forall s \in S : |V^n(s) - V^{n-1}(s)| \leq \varepsilon$. Moreover, a common approach to speeding up convergence comprises updating the estimated values $V^n(s)$ by use of the values previously updated within the same iteration. The resulting so called Gauss-Seidel variant of value iteration[3] uses Update 3.2 as a substitute for Update 3.1.

$$V^n(s) = \max_{d \in D(s)} \left[c(s,d) + \mathbb{E}\left[V^n(s')|s' < s\right] + \mathbb{E}\left[V^{n-1}(s')|s' \geq s\right]\right] \qquad (3.2)$$

Of course, early termination and Gauss-Seidel updates provide only a small contribution to making value iteration a computationally feasible approach to realistic dynamic decision problems. Nevertheless these techniques are the starting point for the simulation based methods introduced in Sect. 3.2.1.

3.1.2 Policy Iteration

Policy iteration (Howard, 1960) is considered as the second elementary method apart from value iteration for solving Bellman's equations. In its basic form a policy is represented as a map $\pi : S \mapsto D$. Alternatively, a policy may be specified as a set of maps with the elements π_t, where $\forall t \in T : \pi_t : S_t \mapsto D_t$. As an example consider the policy resulting from solution of Problem 2.8 for arbitrary $s \in S$. This policy $\pi^\star$ assigns a perfect anticipatory decision to each of the states $s \in S$. The required values $V_t(s_t)$ may be obtained from value iteration. They represent the value of a state subject to perfect anticipatory decisions within the subsequent decision process, i.e. subject to continuous application of $\pi^\star$. Thus $V_t(s_t) = V_t^{\pi^\star}(s_t)$.

Policy iteration iteratively generates a sequence of policies π^n along with the corresponding state values $V^{\pi^n}(s)$. $V^{\pi^n}(s)$ is the value of a state subject to subsequent decisions according to π^n. Policy iteration converges to $\pi^\star$ and to the associated values $V^{\pi^\star}(s)$ respectively within a finite number of iterations. Subsequent iterations are characterized by $\forall s \in S : V^{\pi^n}(s) \geq V^{\pi^{n-1}}(s)$. As a precondition a policy π^0, leading to the terminal state after a finite number of steps, must be present.[4] Procedure 3 gives an outline of policy iteration. A single iteration consists of the two fundamental steps of policy evaluation and policy improvement. In a policy evaluation step, the values $V^{\pi^n}(s)$ corresponding to the current policy π^n are determined. To this end, a system of linear equations with the unknowns $V^{\pi^n}(s)$ must be solved. The values are subsequently used to generate an improved policy π^{n+1} (policy improvement). Convergence occurs as soon as two subsequent policies are identical.

Policy iteration is often preferred to value iteration because a policy of improved qualitiy with respect to the initial policy is guaranteed also in the case of early

[3] See Bertsekas (2007) for proofs of convergence.

[4] Such a policy is often referred to as a "proper policy" in the literature. A detailed elaboration of the concept of a proper policy is provided by Bertsekas (2007, Sect. 2.1).

Procedure 3: Policy Iteration

$n \leftarrow 0$
set π^0
repeat
  solve the system of equations:
  $\forall s \in S : V^{\pi^n}(s) \leftarrow c(s, \pi^n(s)) + \mathbb{E}\left[V^{\pi^n}(s')|s\right]$
  for $(s \leftarrow 1, s \leq k, s{+}{+})$ **do**
    $\pi^{n+1}(s) \leftarrow \arg\max_{d \in D(s)} \left[c(s,d) + \mathbb{E}\left[V^{\pi^n}(s')|s\right]\right]$
  end
  $n{+}{+}$
until $(\pi^{n-1} = \pi^n)$

termination. In addition policy iteration guarantees convergence within a relatively small number of iterations. However, a single iteration requires solution of a system of linear equations, and thus causes much more computational costs compared to a single iteration of the value iteration procedure.

3.1.3 Modified Policy Iteration

Puterman and Shin (1978) show that the values $V(s)$ can be derived by policy iteration without executing exact policy evaluation steps. The corresponding modified policy iteration procedure represents a generalization of both value and policy iteration as introduced in Sects. 3.1.2 and 3.1.1. An outline of modified policy iteration is provided as Procedure 4.

Procedure 4: Modified Policy Iteration

$n, m_n \leftarrow 0$
$\forall s \in S : V_0^{\pi^0}(s) \leftarrow 0$
repeat
  $n{+}{+}$
  for $(s \leftarrow 1, s < k, s{+}{+})$ **do**
    $\pi^n(s) \leftarrow \arg\max_{d \in D(s)} \left[c(s,d) + \mathbb{E}\left[V_{m_{n-1}}^{\pi^{n-1}}(s')|s\right]\right]$
    $V_0^{\pi^n}(s) \leftarrow c(s, \pi^n(s)) + \mathbb{E}\left[V_{m_{n-1}}^{\pi^{n-1}}(s')|s\right]$
  end
  set m_n
  for $(h \leftarrow 1, h \leq m_n, h{+}{+})$ **do**
    $\forall s \in S : V_h^{\pi^n}(s) \leftarrow c(s, \pi^n(s)) + \mathbb{E}\left[V_{h-1}^{\pi^n}(s')|s\right]$
  end
until $(\forall s \in S : V_{m_n}^{\pi^n}(s) = V_{m_{n-1}}^{\pi^{n-1}}(s))$

The procedure generates one policy π^n per iteration by execution of a policy improvement step. In contrast to policy iteration, the values $V_{m_{n-1}}^{\pi^{n-1}}(s)$ used for the improvement step are only estimates of the true values $V^{\pi^{n-1}}(s)$ of policy π^{n-1}. For the first iteration $n = 1$ the estimates used for policy improvement are set in the initialization phase of the procedure. In later iterations, the estimates result from an iterative variant of policy evaluation where the number m_n of evaluation iterations determines the estimates' quality in iteration n. The number m_n may be fixed for all iterations. Alternatively it may either be chosen according to a predefined pattern or selected adaptively depending on the current estimates. Puterman (2005, Sect. 6.5.2) proves modified policy iteration to converge for any of these alternatives.

In case the number m_n of evaluation steps in each iteration n is large enough to guarantee $V_{m_{n-1}}^{\pi^n} = V_{m_n}^{\pi^n}$ the whole procedure coincides with policy iteration. On the contrary, the procedure is identical to value iteration if $m_n = 0$ in each iteration. As in the context of the original value iteration procedure, the criterion for termination of modified policy iteration is based on a threshold value ε in most practical applications (see Sect. 3.1.1).

3.1.4 Linear Programming

Aside from the methods summarized under the umbrella of modified policy iteration, linear programming may be applied for determination of the values $V(s)$. It can be shown (Bertsekas, 2007, Sect. 1.3.4) that the values result from solution of the linear program

$$\begin{aligned} &\max_V \sum_{s=1}^{k} V(s) \\ &\text{subject to} \\ &V(s) \geq c(s,d) + \mathbb{E}\left[V(s')|s\right] \quad \forall s \in S, \forall d \in D(s). \end{aligned}$$

Note that the program comprises one constraint for each state/decision combination. Solving dynamic decision problems appearing in an operational context typically results in linear programs of remarkable size. In general linear programs comprising tens of thousands of variables and constraints may be solved by state of the art algorithms. However, a dynamic decision problem represented by a linear program of this size would be considered as exceptionally small. Despite linear programming is an elegant approach that does not raise the question of convergence, this approach can hardly be implemented for most dynamic decision problems. A more general discussion of the computational limitations is provided in Sect. 3.4.

3.2 Forward Dynamic Programming

The methods of forward dynamic programming are derived from the iterative methods introduced in Sect. 3.1. The term "forward dynamic programming" results from the fact that these methods implement the concepts of dynamic programming by (forward) simulation. In contrast to the idea of backward induction having the terminal state $s_{\mathcal{T}}$ as the basic point of reference, simulation within forward dynamic programming starts from some initial state and then generates a trajectory by moving forward in time through the state space.

The key features of forward dynamic programming are introduced in Sects. 3.2.1 through 3.2.3. In Sect. 3.2.4 these features are integrated into a framework providing a family of forward dynamic programming algorithms. For an extensive discussion of forward dynamic programming we refer to Sutton and Barto (1998) as well as Bertsekas and Tsitsiklis (1996).

3.2.1 Asynchronous State Sampling

The iterative methods of the preceding section basically require looping over all the states $s \in S$ in each iteration. The estimated state values are updated synchronously in the sense that the next update of a particular state is executed only after all the other states received one update in the meantime. Consequently, an update exclusively resorts to values of successor states as derived in the previous iteration. The only exception to this principle is made by the Gauss-Seidel variant of value iteration introduced in Sect. 3.1.1. Gauss-Seidel value iteration resorts to values of states that received an update in the same iteration. This method can be further generalized by allowing for fully asynchronous value updates. In this case a coordination of the frequencies of state value updates is not required any more. In general the resulting asynchronous value iteration converges if every state is updated infinitely often (Bertsekas and Tsitsiklis, 1989).

Asynchronous value iteration requires a sampling mechanism for generating a sequence of states to be updated. For example the sequence might be generated from a probability distribution assigning a fixed update probability to each state. Alternatively the sequence may be generated by simulation of the underlying decision process. A common variant of simulation based state sampling implies making a decision $d^{\star,n} \in D(s)$ according to

$$d^{\star,n} = \arg\max_{d \in D(s)} \left[c(s,d) + \mathbb{E}\left[V^{n-1}(s')|s\right]\right], \tag{3.3}$$

and selecting the next state to be updated according to the transition probabilities $p(s'|s,d^{\star,n})$. This approach is often referred to as real-time dynamic programming in the literature. Conditions for convergence of real-time dynamic programming are due to Barto et al (1993).

Procedure 5: Simulation based asynchronous value iteration

$n \leftarrow 0$
$\forall s \in S : V^0(s) \leftarrow 0$
repeat
  $n{+}{+} \quad t \leftarrow 0$
  randomly select $s_0 \in S_0$
  for $(t \leftarrow 0; t < \mathcal{T}; t{+}{+})$ **do**
    if *(condition for next decision)* **then**
      $d^{\star,n} \leftarrow \arg\max_{d \in D_t(s_t)} \left[c(s_t,d) + \mathbb{E}[V_{t'}^{n-1}(s_{t'})|s_t]\right]$
      $V_t^n(s_t) \leftarrow c(s_t,d^\star) + \mathbb{E}\left[V_{t'}^{n-1}(s_{t'})|s_t\right]$
    end
    $s_{t+1} \leftarrow F(s_t,d^{\star,n})$
  end
until $(\forall s \in S : V^n(s) = V^{n-1}(s))$

Procedure 5 shows a real-time dynamic programming approach as an example for asynchronous value iteration. Note the explicit reference to the natural partitioning of the state space induced by time. The time index of a state $s_t \in S_t$ supports the illustration of the influence of simulation. In each iteration of the procedure an initial state $s_0 \in S_0$ is selected. Starting from s_0 a state trajectory is generated successively. The final state of the trajectory is the terminal state $s_{\mathcal{T}}$ reached at the end of the time horizon. In Procedure 5 a forward move in time resulting in the next state is represented as a stochastic function $F(s_t,d)$. F returns the system state s_{t+1} one time unit ahead of the current time t. State s_{t+1} results from simulation of the exogenous influence subject to the current state s_t as well as subject to the implementation of the previous decision.

Up to here the concept of asynchronous state sampling has been illustrated in the context of value iteration. However, the concept is not limited to value iteration. It is brought into the broader context of policy iteration and modified policy iteration in Sect. 3.2.4.

3.2.2 Monte Carlo Updates

In the previous section simulation is introduced as a means for generating a sequence of states. However, further opportunities exist for using simulation in the context of dynamic programming. In particular, simulated state trajectories can be used for policy evaluation. The resulting procedures are Monte Carlo variants of the iterative way of policy evaluation appearing within modified policy iteration if the values of m_n in Procedure 4 are large (cf. Sect. 3.1.3).

Two basic approaches to policy evaluation by simulation are derived in this section. These approaches update the state values of a policy by Monte Carlo samples generated from simulation. They prove to be special cases of the generic methodology introduced in Sect. 3.2.3. In Sect. 3.2.4 the concept of simulation based Monte Carlo updates is transferred from policy iteration to general forward dynamic programming.

Policy Evaluation by Simulation

Simulation of a trajectory $(s_0, s_{\tau_1}, \ldots, s_{\mathcal{T}})$ subject to a given policy π generates a number of accumulated contributions $C_t(s_t)$ each occuring between a state s_t at time t of the trajectory and the terminal state $s_{\mathcal{T}}$ where

$$C_t(s_t) = c_t(s_t, \pi_t(s_t)) + c_{t'}(s_{t'}, \pi_{t'}(s_{t'})) + \ldots + c_{\tau_{|\tau|}}(s_{\tau_{|\tau|}}, \pi_{\tau_{|\tau|}}(s_{\tau_{|\tau|}})) \,. \tag{3.4}$$

Note that a state s_t may occur at most once at a specific point in time t within a single trajectory. Simulation of N independent trajectories yields a number M of occurences of s_t at t with $0 \leq M \leq N$. The accumulated contribution $C_t^m(s_t)$ derived from the trajectory comprising the mth occurence of s_t at t is a Monte Carlo sample of the true value $V_t^\pi(s_t)$ and $V_t^\pi(s_t) = \mathbb{E}\left[C_t^m(s_t)\right]$. An estimate $\hat{V}_t^{\pi,N}(s_t)$ of $V_t^\pi(s_t)$ results from the sample mean of the accumulated contributions for $m = 1, \ldots, M$, i.e.

$$\hat{V}_t^{\pi,N}(s_t) = \frac{1}{M} \sum_{m=1}^{M} C_t^m(s_t) \,. \tag{3.5}$$

In general, an estimate $\hat{V}_t^{\pi,N}(s_t)$ cannot be determined according to Eq. 3.5 until the whole set of N trajectories is simulated. As an alternative, $\hat{V}_t^{\pi,N}(s_t)$ can be calculated as a moving average. The estimate is then updated iteratively with an iteration occuring right after one single trajectory is simulated. Starting from $\hat{V}_t^{\pi,0}(s_t) = 0$ a new estimate is generated according to Eq. 3.6 if state s_t occurs at time t within trajectory n. $\hat{V}_t^{\pi,n}(s_t) = \hat{V}_t^{\pi,n-1}(s_t)$ if trajectory n does not comprise s_t at t.

$$\hat{V}_t^{\pi,n}(s_t) = \hat{V}_t^{\pi,n-1}(s_t) + \frac{1}{m}\left[C_t^m(s_t) - \hat{V}_t^{\pi,n-1}(s_t)\right] \tag{3.6}$$

Procedure 6 uses iterative value function updates for policy evaluation. Exact policy evaluation requires the number N of simulated trajectories to ensure that each state is visited sufficiently often in order to satisfy $\forall s \in S : \hat{V}^{\pi,N}(s) = V^\pi(s)$. In other words, convergence to the true values $V^\pi(s)$ occurs as $\forall s \in S : m_s \longrightarrow \infty$. An iteration n of policy evaluation by simulation comprises two basic steps. First a trajectory is simulated according to a policy π. Subsequently both the accumulated contribution $C_t(s_t)$ and the updated estimate $\hat{V}_t^{\pi,n}(s_t)$ are calculated for each state s_t included in the trajectory.

This procedure still requires simulation of an entire trajectory before the estimated value function is updated. However, earlier updates are possible by relying on

Procedure 6: Evaluation of a policy π by simulation

```
n ← 0  ∀t∀s_t ∈ S_t : V̂_t^{π,0}(s_t) ← 0 ∧ m_{s_t} ← 0
repeat
    n++  k ← 0
    randomly select s_0 ∈ S_0
    for (t ← 0; t < 𝒯; t++) do
        if (condition for next decision) then
            d_t^⋆ ← π_t(s_t)
            τ_k ← t  k++
        end
        s_{t+1} ← F(s_t, d_{τ_k}^⋆)
    end
    ∀t : V̂_t^{π,n} ← V̂_t^{π,n−1}
    for (i ← k; i ≥ 0; i−−) do
        t ← τ_i  t' ← τ_{i+1}  m_{s_t}++
        C_t(s_t) ← c_t(s_t, d_t^⋆) + C_{t'}(s_{t'})
        V̂_t^{π,n}(s_t) ← V̂_t^{π,n−1}(s_t) + (1/m_{s_t}) [C_t(s_t) − V̂_t^{π,n−1}(s_t)]
    end
until (n = N)
```

the concept of temporal differences. The following paragraph shows that temporal differences allow for an update already after simulation of a single state transition.

Policy Evaluation by Simulation and Temporal Differences

A reformulation of Eq. 3.6 using the single step contributions as well as writing γ instead of $\frac{1}{m}$ results in

$$\hat{V}_t^{\pi,n}(s_t) = \hat{V}_t^{\pi,n-1}(s_t) + \gamma\left[c_t(s_t,d_t^\star) + c_{t'}(s_{t'},d_{t'}^\star) + \ldots + c_{|\tau|}(s_{|\tau|},d_{|\tau|}^\star) - \hat{V}_t^{\pi,n-1}(s_t)\right] .$$

With $V_{\mathcal{T}}(s_{\mathcal{T}}) = 0$ and writing c_t instead of $c_t(s_t,d_t^\star)$ this expression is equivalent to

$$\begin{aligned}\hat{V}_t^{\pi,n}(s_t) = \hat{V}_t^{\pi,n-1}(s_t) + \gamma\Big[&\big(c_t + \hat{V}_{t'}^{\pi,n-1}(s_{t'}) - \hat{V}_t^{\pi,n-1}(s_t)\big)\\ &+ \big(c_{t'} + \hat{V}_{t''}^{\pi,n-1}(s_{t''}) - \hat{V}_{t'}^{\pi,n-1}(s_{t'})\big)\\ &+ \ldots\\ &+ \big(c_{|\tau|} + V_{\mathcal{T}}(s_{\mathcal{T}}) - \hat{V}_{|\tau|}^{\pi,n-1}(s_{|\tau|})\big)\Big] .\end{aligned}$$

A temporal difference δ_t obtained from trajectory n is then defined as the difference between two alternative estimates of $V_t^\pi(s_t)$. Besides $\hat{V}_t^{\pi,n-1}(s_t)$, a temporal difference considers the sum of the contribution received from the current transition and the current estimate of the value of the successor state as an estimate. The quantity

$$\delta_t = \left[c_t + \hat{V}_{t'}^{\pi,n-1}(s_{t'})\right] - \hat{V}_t^{\pi,n-1}(s_t)$$

becomes known immediately after the state transition from s_t to $s_{t'}$. It gives an indication on whether the current estimate of the value of state s_t should be raised or lowered. Considering temporal differences, Eq. 3.6 can finally be reformulated as

$$\hat{V}_t^{\pi,n}(s_t) = \hat{V}_t^{\pi,n-1}(s_t) + \gamma\left[\delta_t + \delta_{t'} + \ldots + \delta_{|\tau|}\right]. \tag{3.7}$$

A single temporal difference δ_t contributes to the value estimate of each of the states $\{s_{\tau_i} | \tau_i \leq t\}$ previously visited within the current trajectory. Executing updates incrementally results in

$$\forall s_{\tau_i} | \tau_i \leq t: \;\; \hat{V}_{\tau_i}^{\pi}(s_{\tau_i}) := \hat{V}_{\tau_i}^{\pi}(s_{\tau_i}) + \gamma\delta_t \tag{3.8}$$

right after the transition from s_t to $s_{t'}$ occured. A valid procedure for policy evaluation requires setting the parameter γ with respect to each update and with respect to each state. According to the principle of moving averages, convergence to the true values $V_t^{\pi}(s_t)$ is guaranteed if γ is set inversely proportional to the number of occurences of the state to be updated. However, a variety of alternative approaches to setting γ exist which are investigated in more detail within Chap. 5.

3.2.3 Stochastic Approximation

The theory of stochastic approximation methods discloses an alternative perspective on policy evaluation by Monte Carlo simulation. It provides a rich family of approaches including those of Sect. 3.2.2 as special cases. In the following, essential stochastic approximation methods for policy evaluation are derived.

Evaluation of a policy π corresponds to solution of a system of equations. On the basis of Sect. 3.1.2 this system can be formulated as

$$\forall t \forall s_t \in S : V_t^{\pi}(s_t) = c_t(s_t, \pi_t(s_t)) + \mathbb{E}\left[V_{t'}^{\pi}(s_{t'}) | s_t\right]. \tag{3.9}$$

However, a number of alternative formulations exist as for example the equations

$$\forall t \forall s_t \in S : V_t^{\pi}(s_t) = \mathbb{E}\left[C_t^m(s_t)\right]. \tag{3.10}$$

The solution of these equations would be straightforward if the expectation could be determined easily. Both the Eqs. 3.9 and 3.10 are an instance of the general h-step Bellman equations

$$\forall \tau_i \forall s_{\tau_i} \in S_{\tau_i} : V_{\tau_i}^{\pi}(s_{\tau_i}) = \mathbb{E}\left[\sum_{k=0}^{h} c_{\tau_{i+k}}(s_{\tau_{i+k}}, \pi_{\tau_{i+k}}(s_{\tau_{i+k}})) + V_{\tau_{i+h+1}}^{\pi}(s_{\tau_{i+h+1}})\right]. \tag{3.11}$$

Independent of the number $h \in \mathbb{N}_0$ selected, a single value $V_t^\pi(s_t) = V_t^\pi$ of a specific state s_t can be obtained from

$$\min_{V_t^\pi} \mathbb{E}\left[\frac{1}{2}(V_t^\pi - Z)^2\right], \tag{3.12}$$

where Z is a random variable with the mean $V_t^\pi(s_t)$. Note that a sample z of Z can be obtained on the basis of any of the h-step Bellman equations. For example following Eqs. 3.10 leads to a sample $z = C_t^m(s_t)$. A solution of Problem 3.12 may be derived by application of a stochastic approximation method.

Stochastic Approximation Methods

The solution of an equation of type $f(x) = x$ can in many cases[5] be determined iteratively. Starting from an initial value x^0 in iteration $n = 0$ and selecting a value $\gamma \in (0, 1]$ a solution $x^\star$ can be sought by repeated calculation of

$$x^{n+1} := (1-\gamma)x^n + \gamma f(x^n). \tag{3.13}$$

This iterative procedure requires the specific function f under consideration to be evaluated exactly. However, in many cases f is unknown. Instead a random variable $W = f(x) + \omega$ may be accessible, with ω being a 0-mean random noise term. Replacing f by W within Procedure 3.13 results in

$$x^{n+1} := (1-\gamma)x^n + \gamma\big(f(x^n) + \omega\big). \tag{3.14}$$

Different flavors of stochastic approximation methods for solution of $f(x) = x$ can be derived from this formulation. The straightforward application of Procedure 3.13 leads to the deterministic iterations

$$x^{n+1} := (1-\gamma)x^n + \gamma \mathbb{E}\left[W\right], \tag{3.15}$$

which however may be hard to carry out due to the need for calculation of the expectation. Instead a number M of random samples w_i of W may be generated, with the sample mean being $\frac{1}{M}\sum_{i=1}^{M} w_i$. A stochastic approximation method is then given by the iterations

$$x^{n+1} := (1-\gamma)x^n + \gamma \frac{1}{M}\sum_{i=1}^{M} w_i. \tag{3.16}$$

Independent of the number M of realizations used for calculating the sample mean, this procedure is a valid means for solution of $f(x) = x$. Note that for large M the sample mean converges to $\mathbb{E}\left[W\right]$. The stochastic approximation method using only a

[5] A detailed description of the required properties of f as well as proofs of convergence are provided by Bertsekas and Tsitsiklis (1996).

single sample per iteration ($M = 1$) was first applied by Robbins and Monro (1951) and is thus often referred to as "Robbins Monro algorithm".

The starting point for solution of Problem 3.12 by means of a Robbins Monro algorithm is given by definition of a function

$$f(V_t^\pi) = V_t^\pi - \mathbb{E}[V_t^\pi - Z] \, .$$

Applying $f(V_t^\pi) = V_t^\pi$ results in $\mathbb{E}[V_t^\pi - Z] = 0$. Note that definition of a function g with $g(V_t^\pi) = \mathbb{E}\left[\frac{1}{2}(V_t^\pi - Z)^2\right]$ implies

$$\mathbb{E}[V_t^\pi - Z] = \nabla g(V_t^\pi) = 0 \, .$$

It is now straightforward to establish a stochastic approximation method with $M = 1$ that leads to the value V_t^π satisfying $\nabla g(V_t^\pi) = 0$. Provided a number of sample realizations z_i the resulting Robbins Monro algorithm is

$$\hat{V}_t^{\pi,n+1} := (1-\gamma)\hat{V}^{\pi,n} + \gamma z_i \, . \tag{3.17}$$

Implementation of this procedure requires a way of generating samples of the random variable Z. A variety of Robbins Monro algorithms results from different ways of generating the values z_i. They are grouped under the umbrella term TD(λ).

Policy Evaluation by TD(λ)

Setting up Robbins Monro procedures for solution of the Eqs. 3.10 results in

$$\forall t \forall s_t \in S_t : \hat{V}_t^{\pi,n+1}(s_t) := \hat{V}_t^{\pi,n}(s_t) + \gamma_{s_t}^n \left(C_t^m(s_t) - \hat{V}_t^{\pi,n}(s_t)\right) , \tag{3.18}$$

with the single sample estimate $C_t^m(s_t)$ being the accumulated contribution received from simulation of a specific trajectory m. Note that policy evaluation with updates according to Eq. 3.6 of the previous section is an instance of the Robbins Monro procedures 3.18.

The Robbins Monro procedure for solution of example System 3.9 requires simulation of only a single transition per iteration. In this example, a single sample estimate of $c_t\left(s_t, \pi_t(s_t)\right) + \mathbb{E}\left[V_{t'}^\pi(s_{t'})|s_t\right]$ is derived from the single step contribution $c_t(s_t, \pi_t(s_t))$ as well as from the current estimate $\hat{V}_{t'}^{\pi,n}(s_{t'})$ of the value of the successor state $s_{t'}$ occuring within the sample trajectory, i.e.,

$$\forall t \forall s_t \in S : \hat{V}_t^{\pi,n+1}(s_t) := \hat{V}_t^{\pi,n}(s_t) + \gamma_{s_t}^n \left[c_t(s_t, \pi_t(s_t)) + \hat{V}_{t'}^{\pi,n}(s_{t'}) - \hat{V}_t^{\pi,n}(s_t)\right] . \tag{3.19}$$

A Robbins Monro procedure can be derived for each h-step Bellman equation. Moreover, a valid Robbins Monro algorithm for policy evaluation can be derived from any weighted average of a number of h-step Bellman equations. Consider the weighted average of all possible Bellman equations as an example. The weighted average is formed by multiplying the Eqs. 3.11 by $(1-\lambda)\lambda^h$ and summing them up over all possible values of h. The family of Robbins Monro procedures derived from the weighted average formulation of Bellman's equations for arbitrary values of $\lambda \in [0,1]$ is widely known as TD(λ) algorithms.[6] During the simulation of a state trajectory, a TD(λ) procedure for determination of the value function of a policy π executes value function updates according to

$$\forall s_{\tau_i} : \hat{V}_{\tau_i}^{\pi,n+1}(s_{\tau_i}) := \hat{V}_{\tau_i}^{\pi,n}(s_{\tau_i}) + \gamma_{s_{\tau_i}}^{n} \sum_{k=i}^{\mathcal{T}-1} \lambda^{k-i} \delta_k \, . \tag{3.20}$$

The parameter λ causes a discounting of the temporal differences δ_k. Hence, the influence of a particular temporal difference on the update of the value of s_{τ_i} depends on the number of transitions between s_{τ_i} and the occurence of δ_k.

For $\lambda = 1$, Procedure 3.20 becomes equivalent to Procedure 3.7, i.e., TD(1) is equivalent to policy evaluation by temporal differences as introduced in Sect. 3.2.2. On the contrary, the TD(0) procedure obtained for $\lambda = 0$ and $0^0 = 1$ coincides with the Robbins Monro procedure 3.19. Every possible TD(λ) algorithm is guaranteed to converge to the values V^{π} with probability 1 provided that both the parameter γ is set appropriately and each state $s \in S$ occurs a sufficient number of times within the simulated state trajectories. Proofs of convergence are due to e.g. Peng (1993), Dayan and Sejnowski (1994) or Tsitsiklis (1994).

A generalization of TD(λ) methods can be achieved by introduction of so called eligibility traces. Augmenting a TD(λ) method with eligibility traces increases the flexibility in discounting temporal differences. In particular, discounting is done by flexible eligibility coefficients instead of the static parameter λ. Eligibility traces are rather a tool for accelerating convergence than a whole new concept. Throughout the years a number of publications on the use of eligibility traces appeared. A proof of convergence of TD(λ) with eligibility traces is provided by Gurvits et al (1994) while a study on the selection of eligibility coefficients is due to Singh and Dayan (1998).

3.2.4 The Actor-Critic Framework

So far stochastic approximation methods were considered as a means of evaluating a policy exactly. However, they also serve for forward dynamic programming based on partial policy evaluation. From a conceptional point of view the resulting methodology is similar to the methodology of Sect. 3.1.3 and represents a generalization of policy iteration by simulation. Methods satisfying this methodology are particularly important with respect to practical application of forward dynamic programming. Unlike strict policy iteration they do not require convergence of a policy evaluation step before policy improvement. For this reason these methods are sometimes referred to as optimistic policy iteration in the literature. Beyond they are

[6] An extensive discussion of TD(λ) including many illustrations is provided by Sutton and Barto (1998).

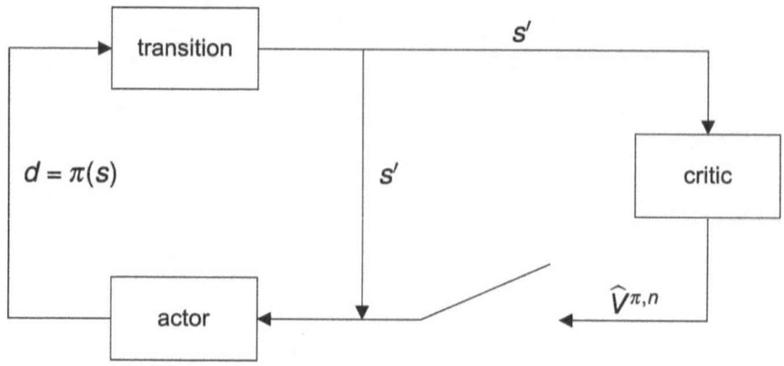

Fig. 3.1 The interaction of actor and critic.

often characterized as actor-critic methods. The notions of "actor" and "critic" reflect the idea of an actor following a policy to make decisions and a critic gradually evaluating the current policy. Figure 3.1 illustrates the interaction between actor and critic.[7] The critic takes into account the state transitions triggered by the actor's decisions and estimates the value function of the actor's policy. From time to time, the critic's current estimates $\hat{V}^{\pi,n}$ are communicated to the actor, causing an update of the policy used for decision making.

A critic may execute policy evaluation by any stochastic approximation method emanating from the previous section. However, in contrast to the methods of Sect. 3.2.3 policy evaluation is subject to early termination. It may be interrupted for improvement of the actor's policy already after having processed a single trajectory or even a single transition only.

Using a TD(λ) algorithm within the actor-critic framework requires selection of both a value of the parameter λ and a frequency of updates of the actor's policy. Procedure 7 shows an actor-critic method with $\lambda = 1$ and policy updates executed every Mth trajectory. A new policy is communicated to the actor a total of N times via the critic's internal value function $\hat{V}$. The actor's current policy results from a policy improvement step based on the values $\hat{V}^{\pi,n-1}$ received from the critic in the previous iteration $n-1$. Note that the application of the policy can be formulated as a policy improvement step due to the contribution function c being deterministic.

Different combinations of λ and the frequency of policy updates lead to different convergence properties. Procedure 7 is guaranteed to converge if $M = 1$ and if the simulated state trajectories are equivalent to synchronous state sampling. It may diverge in the asynchronous case. If a policy update occurs only after a fairly large number of transitions allowing for convergence of the estimates $\hat{V}^{n,m}$ to the true values of policy π^n, each possible actor-critic method is similar to one of the policy iteration methods of the previous sections. Updating after each single transition and setting $\lambda = 0$ leads to the asynchronous value iteration of Procedure 5. Both the policy iteration methods of the previous sections and asynchronous value iteration converge to the values of the optimal policy $\pi^\star$.

[7] The figure is in the style of Fig. 5.3 in Bertsekas and Tsitsiklis (1996).

Procedure 7: An actor-critic method using TD(1) for policy evaluation.

$n \leftarrow 0 \;\; \forall t \forall s_t \in S_t : \bar{V}_t^{\pi^0}(s_t) \leftarrow 0$
repeat
  $n{+}{+} \;\; m \leftarrow 0 \;\; \forall t \forall s_t \in S_t : \hat{V}_t^{n,0}(s_t) \leftarrow \bar{V}_t^{\pi^{n-1}}(s_t)$
  repeat
    randomly select $s_0 \in S_0 \;\; k \leftarrow 0 \;\; m{+}{+}$
    for $(t \leftarrow 0; t < \mathscr{T}; t{+}{+})$ **do**
      if *(condition for next decision)* **then**
        $d_t^{\star,n,m} \leftarrow \arg\max_{d \in D_t(s_t)} \left[c_t(s_t,d) + \mathbb{E}[\bar{V}_{t'}^{\pi^{n-1}}(s_{t'})|s_t]\right]$
        $\tau_k \leftarrow t \;\; k{+}{+}$
      end
      $s_{t+1} \leftarrow F(s_t, d_{\tau_k}^{\star,n,m})$
    end
    $\forall t \forall s_t \in S_t : \hat{V}_t^{n,m}(s_t) \leftarrow \hat{V}_t^{n,m-1}(s_t)$
    for $(i \leftarrow k; i \geq 0; i{-}{-})$ **do**
      $t \leftarrow \tau_i \;\; t' \leftarrow \tau_{i+1}$
      $C_t(s_t) \leftarrow c_t(s_t, d_t^{\star,n,m}) + C_{t'}(s_{t'})$
      $\hat{V}_t^{n,m}(s_t) \leftarrow \hat{V}_t^{n,m-1}(s_t) + \gamma[C_t(s_t) - \hat{V}_t^{n,m-1}(s_t)]$
    end
  until $(m = M)$
  $\forall t \forall s_t \in S_t : \bar{V}_t^{\pi^n}(s_t) \leftarrow \hat{V}_t^{n,M}(s_t)$
until $(n = N)$

A more detailed discussion of the convergence properties of actor-critic methods is provided by Bertsekas and Tsitsiklis (1996, Sect. 5.4). However, the convergence properties are not fully understood until today (Bertsekas, 2007, Sect. 6.2.3). A number of actor-critic methods do not guarantee perfect anticipation. Nevertheless the theoretical properties of actor-critic methods are not always correlated to their practical success. Thus, both the non-convergent methods and the early termination variants of convergent methods may be considered for anticipatory optimization of dynamic optimization problems.

3.3 Model Free Dynamic Programming

Both the dynamic programming approaches of Sect. 3.1 and the forward dynamic programming methods of Sect. 3.2 require the transition probabilities $p_t(s_{t'}|d_t, s_t)$ to be available. That is to say they rely on a model of the exogenous process influencing state transitions. The methods under the umbrella of modified policy iteration make use of transition probabilities for the purpose of both policy evaluation and policy improvement. By contrast, the simulation based actor-critic framework evaluates a policy without a model of the exogenous process but still requires transition probabilities for the purpose of policy updates. In particular the values $p_t(s_{t'}|d_t, s_t)$ are required for evaluation of the expectation within expressions of type

$$d_t^\star = \arg\max_{d \in D_t} \left[c_t(s_t, d) + \mathbb{E}\left[\hat{V}_{t'}^{\pi}(s_{t'}) | s_t \right] \right] .$$

If a model of the exogenous process is unknown, the evaluation of the expectation could be done by means of estimated probabilities, possibly derived as mean values from repeated simulation of state transitions. The major disadvantages of such an approach would be both the presumable inaccuracy of the estimates and the huge amount of additional computational burden due to the additional simulation of state transitions. Nevertheless, even if transition probabilities are available, application of the methods of the previous sections may cause difficulties. The combination of a huge number of relevant successor states and the possibly very small values p_t is very likely to result in considerable numerical instability in the evaluation of an expectation.

In the following two variants of perfect anticipation are provided that both do not rely on a model of the exogenous process. Transition probabilities are not required. The variant illustrated in Sect. 3.3.1 can be applied in the presence of any type of interaction of a decision and the random influence on a state transition. Section 3.3.2 contains a variant of perfect anticipation for the case of separability (cf. Sect. 2.1.1) of the influences of a decision and the exogenous process on a state transition.

3.3.1 Q-Learning

The most popular variant of model free dynamic programming deals with the values of state-decision pairs instead of the values of states only. For a given policy π the value of an arbitrary pair (s_t, d_t) at time t results from the value function

$$Q_t^{\pi}(s_t, d_t) = \left[c_t(s_t, d_t) + \mathbb{E}\left[V^{\pi}(s_{t'}) | s_t \right] \right] . \tag{3.21}$$

$Q_t^{\pi}(s_t, d_t)$ can be considered as a value function on an extended state space established by the combination of the states $s_t \in S_t$ with each of the decisions $d_t \in D_t(s_t)$. In accordance with the original works on model free dynamic programming (Watkins, 1989) a particular value of $Q_t^{\pi}(s_t, d_t)$ is denoted as a Q-Factor and the methods incorporating Q-Factors are referred to as Q-Learning.

An optimal Q-Factor $Q_t^{\pi^\star}(s_t, d_t)$ represents the value of taking decision $d_t \in D_t(s_t)$ in state s_t at time t and following the optimal policy $\pi^\star$ subsequently. Reformulating Bellman's equation (c.f. Eq. 2.6) by means of the optimal Q-Factors leads to

$$V_t^{\pi^\star}(s_t) = \max_{d_t \in D_t(s_t)} Q_t^{\pi^\star}(s_t, d_t) . \tag{3.22}$$

Eqs. 3.21 and 3.22 imply the values $Q_t^{\pi^\star}(s_t, d_t)$ to be the solution of the system of equations

$$\forall(s_t,d_t) : Q_t^{\pi^\star}(s_t,d_t) = \left[c_t(s_t,d_t) + \mathbb{E}\big[\max_{d_{t'} \in D_{t'}(s_{t'})} Q_{t'}^{\pi^\star}(s_{t'},d_{t'})|s_t\big]\right] . \tag{3.23}$$

This system may be solved by use of the stochastic approximation methods introduced in Sect. 3.2.3. If a simulation environment for the generation of successor states $s_{t'}$ is provided, the Q-Factors may be sought by use of the Robbins Monro method

$$\hat{Q}_t^{n+1}(s_t,d_t) := (1-\gamma)\hat{Q}_t^n(s_t,d_t) + \gamma\big[c_t(s_t,d_t) + \max_{d_{t'} \in D_{t'}(s_{t'})} \hat{Q}_{t'}^n(s_{t'},d_{t'})\big] . \tag{3.24}$$

The resulting approach shares many similarities with an actor-critic method featuring $M = 1$ and $\lambda = 0$. The most important difference between the two, apart from the fact that Q-Factors are estimated instead of a value function $V^{\pi^\star}(s)$, consists in the way a decision is made. While an actor-critic method needs to evaluate the expectation of the value of the successor state, Q-Learning simply derives a decision by solution of

$$d_t^\star = \arg \max_{d_t \in D_t(s_t)} \hat{Q}_t^{n-1}(s_t,d_t) . \tag{3.25}$$

Watkins and Dayan (1992) prove that Q-Learning by means of Update 3.24 converges to the optimal Q-Factors. Note that in the presence of optimal Q-Factors an anitcipatory decision no longer requires solution of a stochastic optimization problem P_t (cf. Problem 2.8). Instead the deterministic problem $P_t = \left(D_t(s_t),\, Q_t^{\pi^\star}(s_t,d_t)\right)$ must be solved each time an anticipatory decision is made. A model of the exogenous process is neither required for learning the value function $Q_t^{\pi^\star}$ nor for the implementation of anticipatory decisions. However, the benefits of Q-Learning are gained at the expense of a more complex value function on the elements of an extended state space.

3.3.2 Post-decision States

An element (s_t,d_t) of the extended state space needed for Q-Learning represents the knowledge about resources and other system attributes at time t right after decision d_t is made. As a decision d_t codetermines the successor state $s_{t'}$ of s_t, the available knowledge about the system increases as soon as d_t is known. A state capturing this knowledge is often referred to as a post-decision state in the literature.

A more compact representation of a post-decision state is possible if a decision's influence on a state transition can be separated from the random influence of an exogenous process. As indicated in Sect. 2.1.1, in such a case the knowledge right after a decision is made may be summarized by means of a post-decision state variable $s_t^d \in S_t^d$ whose complexity does not exceed the complexity of the regular (pre-decision) state variable $s_t \in S_t$.

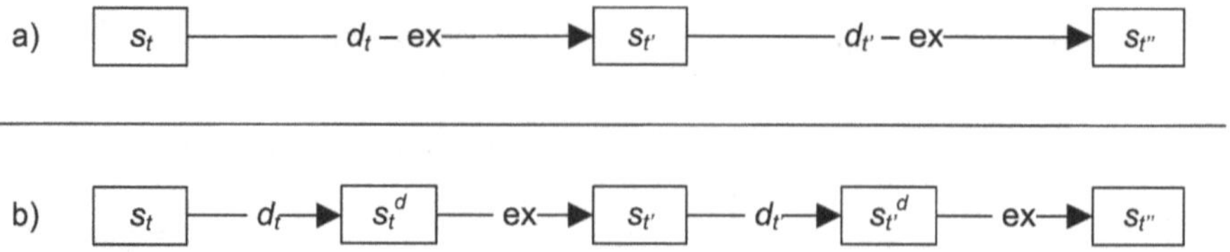

Fig. 3.2 Representation of a system trajectory *a)* without post-decision states and *b)* showing both post- and pre-decision states.

Figure 3.2 illustrates the characteristics of this type of post-decision state variable. The figure's parts a) and b) show two different representations of one state trajectory. Part a) gives a perspective on the trajectory involving pre-decision states only. On the contrary, part b) shows both the pre-decision and the post-decision states occuring within the trajectory. Part a) unveils the fact that a transition is determined by both a decision and a - possibly intertwined - exogenous random influence, denoted as "ex". More details are provided in part b) showing that the system moves into a post-decision state immediately after a decision is made. A post-decision state captures the system knowledge with the current decision being applied. It becomes known at the same time the decision is made.

Figure 3.2 suggests that the value function V_t^d around the post-decision states $s_t^d \in S_t^d$ may be formulated subject to the value functions $V_{t'}$ of the subsequent pre-decision states $s_{t'} \in S_{t'}$, i.e.

$$V_t^d(s_t^d) = \mathbb{E}[V_{t'}(s_{t'})|s_t] \ . \tag{3.26}$$

Moreover the value function around the post-decision state variables allows for the reformulation of Bellman's equations as

$$V_{t'}(s_{t'}) = \max_{d_{t'} \in D_{t'}} \left[c_{t'}(s_{t'}, d_{t'}) + V_{t'}^d(s_{t'}^d)\right] \ . \tag{3.27}$$

Substituting Eq. 3.27 into Eq. 3.26 leads to Bellman's equation around the post-decision state variables

$$V_t^d(s_t^d) = \mathbb{E}\left[\max_{d_{t'} \in D_{t'}} \left[c_{t'}(s_{t'}, d_{t'}) + V_{t'}^d(s_{t'}^d)\right] |s_t^d\right] \ . \tag{3.28}$$

The key difference between this formulation and the Bellman equations around the pre-decision state variables consists in the role of the expectation. The maximization operator within the expectation has a significant effect on the forward dynamic programming methods of the previous section. An actor-critic method for determination of the value function on the post-decision states is represented as Procedure 8. The benefit from using the value function around the post-decision states becomes apparent in the decision making step. In contrast to the actor-critic method of Procedure 7 the optimization problem to be solved is deterministic and the state transition probabilities are not required for solution.

When it comes to the estimation of the value function on the post-decision states, the procedure exploits the fact that a sample estimate of the value of a pre-decision state is at the same time a sample estimate of the value of the preceding post-decision state. In the case of TD(1) an accumulated contribution $C_{\tau_i}(s_{\tau_i})$ is a sample estimate of the value $V^d_{\tau_{i-1}}(s^d_{\tau_{i-1}})$ of the previous post-decision state $s^d_{\tau_{i-1}}$.

Procedure 8: An actor-critic method relying on post-decision state variables.

$n \leftarrow 0 \quad \forall t \forall s^d_t \in S^d_t : \bar{V}^{d,\pi^n}_t(s^d_t) \leftarrow 0$
repeat
  $n{+}{+} \quad m \leftarrow 0 \quad \hat{V}^{d,n,0} \leftarrow \bar{V}^{d,\pi^{n-1}}$
  repeat
    randomly select $s_0 \in S_0 \quad k \leftarrow 0 \quad m{+}{+}$
    for $(t \leftarrow 0; t < \mathscr{T}; t{+}{+})$ **do**
      if *(condition for next decision)* **then**
        $d^{\star,n,m}_t \leftarrow \arg\max_{d \in D_t(s_t)} \left[c_t(s_t,d) + \bar{V}^{d,\pi^{n-1}}_t(s^d_t)\right]$
        $\tau_k \leftarrow t \quad k{+}{+}$
      end
      $s_{t+1} \leftarrow F(s_t, d^{\star,n,m}_{\tau_k})$
    end
    $\forall t \forall s^d_t \in S^d_t : \hat{V}^{d,n,m}_t(s^d_t) \leftarrow \hat{V}^{d,n,m-1}_t(s^d_t)$
    for $(i \leftarrow k; i \geq 0; i{-}{-})$ **do**
      $t \leftarrow \tau_i \quad t' \leftarrow \tau_{i+1} \quad t^\dagger \leftarrow \tau_{i-1}$
      $C_t(s_t) \leftarrow c_t(s_t, d^{\star,n,m}_t) + C_{t'}(s_{t'})$
      $\hat{V}^{d,n,m}_{t^\dagger}(s^d_{t^\dagger}) \leftarrow \hat{V}^{d,n,m-1}_{t^\dagger}(s^d_{t^\dagger}) + \gamma\left[C_t(s_t) - \hat{V}^{d,n,m-1}_{t^\dagger}(s^d_{t^\dagger})\right]$
    end
  until $(m = M)$
  $\forall t \forall s^d_t \in S^d_t : \bar{V}^{d,\pi^n}_t(s^d_t) \leftarrow \hat{V}^{d,n,m}_t(s^d_t)$
until $(n = N)$

3.4 Limited Effectiveness of Perfect Anticipation

Sections 3.1 through 3.3 provide the means for perfect anticipation. With the exception of some instances of the actor-critic framework of Sect. 3.2.4, all of the introduced methods are guaranteed to converge to the actual value function at least with probability 1.

Nevertheless the actual realization of perfect anticipation may be hard. Applying the methods of the previous sections implies a prohibitive computational effort for the majority of the dynamic decision problems arising from an operational context. On the one hand, the mere number of iterations required for convergence may be huge. On the other hand the problems' structure may lead to a tremendous amount of computation already within one single iteration. Last but not least the methods for perfect anticipation imply significant memory requirements. Keeping track of the value function for each state may be a task as prohibitive as the merely computational challenges. In particular, the limited effectiveness of perfect anticipation originates from three sources:

- **State space dimensionality.** The dimensionality of the state space is determined by the number $\mathcal{T}$ of time periods considered as well as by the attributes required for representation of the relevant system knowledge. Both the number of attributes and the number of possible values of each of the attributes have an impact. The resulting state space may be vast.
  The state space dimensionality is a major determinant of the computational effort with respect to each of the methods of Sections 3.1 through 3.3. As an example the value iteration procedure of Sect. 3.1.1 comprises looping over the whole state space in each iteration. A modified policy iteration approach as introduced in Sect. 3.1.3 even requires looping over the whole state space $(m_n + 1)$ times per iteration. Moreover, the simulation based methods generally call for visiting each state a (possibly large) number of times, in order to guarantee convergence. On top of that, all the methods must retain in memory the single values of the whole set of states.
- **Decision space complexity.** Every approach to perfect anticipation implies repeated determination of a decision maximizing the total of contribution and estimated value of successor states. Determining such a decision raises an optimization problem whose complexity heavily depends on both, the nature of the contribution function c and the structure of the set of possible decisions. The latter results from the number of attributes required for representation of a single decision as well as from the constraints on the attributes' possible values. Dynamic decision problems in an economic context typically comprise various interdependences of attribute values leading to optimization problems of considerable complexity. In particular such an optimization problem may belong to the class of NP-hard problems, which means that the problem may not be solvable in polynomial time. Note that there may be the need to solve hundreds, thousands or even millions of such optimization problems in order to treat a dynamic decision problem.
- **Expected values.** The need for evaluation of expected values provides additional challenges within the methods for perfect anticipation. These challenges motivate the model free approaches of Sect. 3.3. However, the model free approaches do not reduce computational difficulties in any case. They do if a decision's impact on the state transition can be separated from the random influence (cf. Sect. 3.3.2). Also in case of inseparability evaluation of expected values can be circumvented. However, the Q-Learning approaches (cf. Sect. 3.3.1) for model free dynamic programming lead to a possibly tremendous increase of the state space dimensionality.

Typically all three sources of limitation occur in a dynamic decision problem. However, the presence of only one of the sources may be sufficient for making the application of the methods of Sects. 3.1 through 3.3 prohibitive. Nevertheless, the dimensionality of the state space is widely considered as being the most capital source of limitation. In the literature it is often referred to as the "curse of dimensionality". As a consequence an increasing number of authors refer to the three sources as the three curses of dimensionality. The curses of dimensionality set limits to the subset of dynamic decision problems allowing for the precise determination of the value function V. In reality V can only be determined for problems of tiny dimensionality. In view of todays computational possibilities perfect anticipation does not take effect for other problems.

However, the curses of dimensionality can be counteracted. The price to pay for counteraction is a lower degree of anticipation. In terms of the prototypical degrees of anticiaption introduced in Sect. 2.3.2 the step from perfect anticipation towards approximate anticipation suggests itself. Yet, the realization of approximate anticipation requires an extended methodology. In particular, approximate anticipation for dynamic decision making must take advantage of the synergies of optimization and data mining.

Chapter 4
Synergies of Optimization and Data Mining

Data mining provides the concepts for reducing the negative effects of a vast state space. Both these concepts and optimization are required for the realization of approximate anticipation. In particular approximate anticipation emerges as a synergy of optimization and data mining. However, these two procedures are basically independent of each other. Thus, the present chapter provides the general foundations of the synergies of optimization and data mining, based on which the next Chap. 5 introduces approximate anticipation. A unified view of the optimization and data mining is established in Sect. 4.1. Subsequently the synergies of the two are both identified and illustrated with respect to their contribution to data mining in Sect. 4.2 and with respect to optimization in Sect. 4.3.

4.1 Preliminaries

Three consecutive steps lead to a unified view of optimization and data mining. Common foundations of the two are established in Sect. 4.1.1. Then a data mining process model is defined and embedded into these foundations in Sect. 4.1.2. Finally an integrated methodology consisting of both data mining and optimization is presented in Sect. 4.1.3.

4.1.1 Common Foundations

Both optimization and data mining are application focused (White, 1991; Wu et al, 2003). Thus, a common point of reference is given by the system the two procedures are applied to. However, such a system is naturally tied to a specific application domain. A generic view of system elements is necessary to understand the general roles of optimization and data mining as well as the synergies of the two.

Each system consists of a set of interrelated objects, with an object being either a real entity or a mere concept. An object comprises a set of attributes, leading to a more precise definition of a system as a relation among attributes of objects (Mesarovic and Takahara, 1989). The value of a single attribute may be either constant or variable. In the context of dynamic decision making an example of the latter is provided by the knowledge about the system represented as the attributes of the state variable.

A system's appearance is given by the values its constituting attributes hold as the system is observed at a particular point in time t. Changing values of single attributes results in a modification of the system's appearance. A sequence of appearances of a system is percieved as system behavior. System behavior is restricted by the domains of the values of the system attributes. In the case of dynamic decision making a system appearance depends on its predecessor via the state transition function. However, system behavior does not generally imply subsequent appearances to depend on each other. The following Example 1 illustrates a system with subsequent appearances being interdependent. A system comprising appearances that are independent of each other is provided as Example 2:

- Example 1 - Fleet surveillance: An example system from the application domain of transportation is defined as consisting of a number of vehicles, a number of potential customers and a number of road links as system objects. The variable system attributes are 'location' and 'load' of each vehicle, the 'quantity demanded' of each customer as well as the 'estimated travel time' associated with each road link. The system is completed by adding the 'geographical location' of each customer as system attributes whose values are constant.
  A similar system is considered by Attanasio et al (2007) for fleet surveillance within a same-day courier service company. The company constantly records system appearance by observing the values of the variable system attributes. System behavior is due to vehicle operations, customer orders and changing road conditions.
- Example 2 - Market basket analysis: A typical system examined in the application domain of marketing consists of a number of products - possibly sold by a retail store - as objects. Each product is assigned the variable attribute 'quantity sold'. Such a system has been considered many times in the literature on market basket analysis (e.g., Agrawal et al, 1993; Brijs et al, 2004). In the standard market basket analysis approach a system appearance is identified with a sales transaction. Hence, subsequent system appearances will usually be to a large degree independent of each other.

As specified in Eq. 4.1, a system $\mathscr{S}$ consisting of a number N of attributes is represented as a subset of the Cartesian product of these attributes' domains. Based on this fundamental system model, a system appearance $\sigma \in \mathscr{S}$ is given by an N-tuple of the current values $\{a_n \in \mathscr{A}_n | n \in \{1 \ldots N\}\}$ of the system attributes.

$$\mathscr{S} \subseteq \mathscr{A}_1 \times \ldots \times \mathscr{A}_N \,. \tag{4.1}$$

A special case occurs if one or more agents deliberately affecting the system are represented as proper system objects. An object corresponding to an agent has two specific types of variable attributes. On the one hand, the agent's decisions on how to affect the system must be considered. On the other hand, attributes representing the agent's evaluation of his decisions are needed. The domain of one of the former is denoted by $\mathscr{D}_m$. The domain of one of the latter attributes is denoted by $\mathscr{E}_k$. The refined model of a system comprising N regular system attributes as well as M decision attributes and K evaluation attributes is denoted as

$$\mathscr{S} \subseteq \mathscr{A}_1 \times \ldots \times \mathscr{A}_N \times \mathscr{D}_1 \times \ldots \times \mathscr{D}_M \times \mathscr{E}_1 \times \ldots \times \mathscr{E}_K \, . \tag{4.2}$$

The structural determinants of a system usually make $\mathscr{S}$ a true subset of the Cartesian product. System structure is identified by means of the following types of relations:

- *Static relations* represent constraints on the values of single attributes as well as interdependences among the values of attributes of an appearance.
- *Dynamic relations* specify the interdependence of the values of (possibly different) attributes with respect to subsequent system appearances. Consider a function characterizing state transitions in dynamic decision making as an example. Dynamic relations do not exist if subsequent appearances are independent of each other.

The following example illustrates the concept of system structure based on the refined system model of Eq. 4.2.

- Example 3 - Fleet management: A natural extension of the fleet surveillance system (Example 1) is due to an agent taking decisions on the operation of the single vehicles within the system. The real-time fleet management system developed by Attanasio et al (2007) provides an example. Decision attributes are introduced for the assignment of customer requests to vehicles and for repositioning the vehicles respectively. The evaluation of these decisions is carried out in terms of courier efficiency and service times. Consequently evaluation attributes are introduced for both of them.
  Examples of static relations: The domain of a single attribute may be bounded by definition. Thus, the value of a vehicle's 'load' may be subject to some minimum and some maximum respectively, where both the minimum value and the maximum value may be a constant system attribute. Similar bounds may be introduced for other attributes. Moreover, the domain of a decision attribute may be restricted by the value of another attribute of the same appearance. Consider the relation between the decision to serve a customer and the attribute 'quantity demanded' of this customer as an example. Typically a customer can be served only in case its 'quantity demanded' is more than zero. In addition, the domain of an evaluation attribute may be restricted by the value of a decision attribute of the same appearance. As an example, courier efficiency is likely to depend on service decisions.

Examples of dynamic relations: The interdependence of attribute values of subsequent appearances may be due to the dynamics inherent in the system objects. For instance, the behavior of a customer implies a dynamic relation between his 'quantity demanded' of subsequent appearances. A similar statement can be made for the traffic dynamics on a road link and the 'estimated travel time' associated with this link. Moreover, a dynamic relation may be provoked by a decision making agent. As an example consider the dynamic influence of the decision to serve a customer now on the customer's 'quantity demanded' of a subsequent appearance.

Observation of system behavior is the elementary way of gaining knowledge about a system. Recording system appearances provides historical data representing this knowledge. However, there may be the desire to know values of system attributes beyond the data actually available from observation. More precisely, it might either not be possible to observe the values of certain attributes of an appearance at hand or the value of an attribute of a future appearance might be unknown. In both of these cases information about system structure is needed for inference of the unknown attribute values:

- Information about static relations allows for drawing conclusions on the value of an attribute based on the values of other attributes of the same system appearance, i.e. it provides a model for inference of facts that already happened but cannot be observed.
- Information about dynamic relations allows for making forecasts about values of attributes of future appearances, i.e. it provides a model for inference of facts that did not happen yet.

The methodology leading to information about system structure is introduced in the next section.

4.1.2 Data Mining

Data mining generates information about system structure. In terms of Sect. 4.1.1 it seeks for a representation of either dynamic or static relations. To this end it is concerned with secondary analysis of a huge amount of data typically being at hand in the form of recorded system appearances. Both the aspect of secondary analysis and the sheer size of data distinguish the field from common statistics (Hand, 1998, 1999). In contrast to common statistics data is not collected based on experiments designed to answer a certain set of a priori known questions. Instead data mining processes data arising as a byproduct of either a system in actual operation or a system simulation. However, the data cannot be processed without at least a vague hypothesis about system structure. Such an a priori hypothesis determines the type of information to be derived from data.

While data mining was discussed critically in its early days (e.g., Lovell, 1983; Denton, 1985), it has soon been recognized and well-established as the scientific

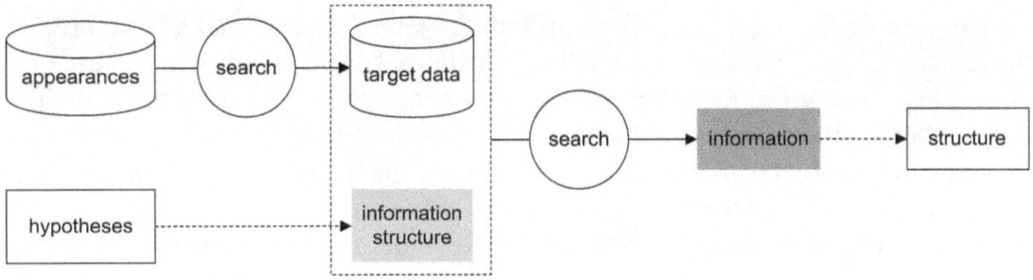

Fig. 4.1 The data mining process model. *Solid line arrows* indicate maps, implemented by search operators. *Dashed line arrows* represent deductions made by the user.

methodology for computer automated analysis of large and complex data sets (e.g., Friedman, 1997). Nevertheless, a unique definition of data mining has not been established yet. Thus a number of process models exist. Virtually each of them considers data mining from a business perspective. One of the most popular approaches is due to Berry and Linoff (2011). They define data mining as a business process and derive a ten step process model from this perspective. A similar idea provides the basis for another popular model named the CRoss-Industry Standard Process for Data Mining (CRISP-DM). For details about the six phases of CRISP-DM see, e.g., Shearer (2000). An example application of the CRISP-DM model within the field of customer relationship management is provided by Li et al (2006).

From a more scientific perspective, the data mining process was shaped by two milestones in data mining evolution. Fayyad et al (1996) established the extraction of information about system structure from recorded system appearances as a multi-step procedure. They distinguish between a preprocessing step and the core data mining activity. Hand et al (2001) elaborate on the latter by specifying a framework for core data mining algorithms.

Meisel and Mattfeld (2007b) describe the transition from both recorded appearances and hypotheses to information about system structure as illustrated in Fig. 4.1. The preprocessing step is denoted as a map from a set of recorded appearances into a target data set. The implementation of this map may comprise various instances of the four types of preprocessing operations:

- Selection of a subset of appearances
- Projection of the system attributes into a subset of these attributes
- Transformation of the values of certain attributes of each appearance, i.e. modification of the domain of an attribute
- Construction of new attributes by aggregation of the original attributes

The significance of preprocessing was emphasized many times in the literature. A recent study (Crone et al, 2006) shows that the core data mining step is almost as sensitive to variations of target data as to different parameterizations of the core algorithm. Typically a preprocessing operation is implemented as a search procedure that should return the optimal target data set with respect to the performance criteria of the subsequent core data mining step.

The core data mining step relies on both the target data set and a structure the information to be derived must obey. The selection of such an information structure is based on the a priori hypotheses about system structure. Typically, the core data mining step is implemented as a search procedure gradually fitting the information structure to the target data set. Hand et al (2000) identify two alternative types of information that may be identified by the search procedure. They distinguish between a *model* representing structure inherent in the whole set $\mathscr{S}$ of appearances and a *pattern* representing structure inherent in only a subset of $\mathscr{S}$. Casually speaking, patterns represent exceptional system behavior while models represent global statements about system behavior. The different types of models and patterns available are categorized according to the following data mining tasks (Hand et al, 2001). For examples and more details about the tasks, we refer to Sect. 4.2.

- *Descriptive Modeling* results in models representing an aggregated view of the set of system appearances. No single attribute is central to a descriptive model. However, such a model may be used for inference of a missing value of an attribute of any single system appearance. Cluster analysis is the most popular among the techniques existing for this data mining task. A thorough discussion of descriptive modeling techniques is provided in Hand et al (2001).
- *Predictive Modeling* results in models representing a relation between one particular system attribute and a set of others. Such models either allow for forecasts of attribute values of future appearances or they allow for inference of an attribute value that could not be observed in the past. Common techniques are for instance decision trees (Quinlan, 1986; Breimann et al, 1984) and support vector machines (Burges, 1998). Fisher's discriminant analysis (Fisher, 1936) is known as the earliest predictive modeling technique.
- *Discovering Patterns and Rules* results in patterns allowing for inference of attribute values of system appearances subject to the presence of specific values of other attributes within the same appearance. In the majority of cases the patterns are represented as rules. Important techniques are association rule algorithms (Agrawal et al, 1993; Agrawal and Srikant, 1994), rule induction methods (Quinlan, 1993, 1986; Breimann et al, 1984) as well as rough set approaches (e.g., Slowinski et al, 2005; Greco et al, 2005; Slowinski et al, 2008). The techniques essentially differ with respect to both the information structure and the search method required.
- *Exploratory Data Analysis* relies on visual or interactive techniques. An analytical model formulation might not be possible in any case. However, a visual or interactive model may provide useful insights into system structure as well support the generation of new hypotheses.

Eventually the information derived by data mining must be verified, i.e. it must be determined to what extent the derived model or pattern actually represents system structure. Verification is one of the most difficult steps within the data mining process. Standard methods rarely exist. In many cases verification is done by comparison of attribute values derived from a model or pattern respectively to attribute values derived from observed system appearances.

4.1.3 Integration of Optimization and Data Mining

Section 4.1.1 lays a common foundation of optimization and data mining from an application perspective. Indeed, many data mining approaches are within traditional application domains of optimization, such as transportation (e.g., Yuan and Cheu, 2003), manufacturing (e.g., Ötztürk et al, 2006), health care (e.g., Li et al, 2004) or finance (e.g., Leigh et al, 2005). As the two procedures are focused on similar types of applications, it is a natural step to strive for synergies. To this end, optimization and data mining should be considered as parts of one integral problem solving methodology (Smith and Gupta, 2000).

Only recently, a number of special issues (Olafsson, 2006; Karasözen et al, 2006; Feng and Kusiak, 2006; Osei-Bryson and Rayward-Smith, 2009; Chen, 2010) of international research journals were dedicated to successful approaches integrating optimization and data mining. However, these approaches either focus on a particular application domain or treat a very specific method only. A general categorization of integrative approaches is provided by Meisel and Mattfeld (2007a). The same authors are the first to define a generic methodology comprising optimization and data mining (Meisel and Mattfeld, 2010). In this context they consider optimization with respect to the common foundations established in Sect. 4.1.1.

From the perspective of a refined system model as specified in Eq. 4.2, optimization determines the values of the decision attributes of a system appearance. Hence, a specific optimization procedure may be considered as imposing system structure by definition of a static relation between decision attributes and other system attributes. At the same time however optimization requires information about system structure as a critical precondition. The process model leading from hypotheses about system structure to the values of decision attributes consists of three major steps:

1. Identification of a basic system model: The model is an instance of Eq. 4.1. It may be directly available from observation of one or more system appearances. If a domain of an attribute cannot be fully observed, hypotheses about system structure must be consulted.
   In the case of dynamic decision making, the variable attributes of the basic system model are equivalent to the attributes of the state variable. If the system has no variable attributes at all, the basic system model comprises a single appearance only, i.e. the domains $\{\mathscr{A}_n | n \in \{1 \dots N\}\}$ contain a single value each.
2. Model refinement by decision attributes and evaluation attributes: Introduction of these additional attributes implies additional system structure. A representation of both the domains $\{\mathscr{D}_m | m \in \{1 \dots M\}\}$ of the decision attributes and the domains $\{\mathscr{E}_k | k \in \{1 \dots K\}\}$ of the evaluation attributes is derived as a decision model. In terms of Sect. 2.2.1, a decision model represents both the set of feasible decisions X and the possible values of the objective function f of an optimization problem $P = (X, f)$.
   Each of the appearances identified in step one implies its own decision model. However, these models are related to the same system and consequently share a

common structure. Thus, a decision model is derived as an instance of a generic decision model structure. Identification of the decision model structure is a crucial step. If the generic structure defining the domains of decision attributes or evaluation attributes cannot be identified easily, hypotheses must be consulted. These hypotheses may have a significant impact on the set of possible search procedures to be applied in step three.
3. Application of a search procedure: The search procedure applied implements a map from the domains of regular system attributes into the domains of decision attributes. It seeks for those values of the decision attributes receiving the best evaluation possible. Basically any search procedure corresponding to the categories of optmization techniques introduced in Sect. 2.2.2 may be considered for application. However, limitations of the set of applicable search procedures are subject to the decision model structure.

Note that both step one and two may heavily depend on hypotheses about system structure. If the hypotheses do not originate from precise information about system structure, the optimization process runs the risk of being pushed into absurdity.

A unified view of optimization and data mining is provided in Fig. 4.2. Both structure and appearance of the system at hand, constitute the common points of reference of the methodology. Optimization derives the values of decision attributes, while data mining identifies information about system structure. However, the figure highlights another important link between optimization and data mining. Both procedures may heavily rely on hypotheses about system structure. As part of the required hypotheses may be generated beyond algorithmic information processing, the dependency on hypotheses also shows the fundamental limitations of optimization and data mining. Ultimately experimentation is needed in order to conduct the procedures. Recently the idea of experimentation as a physical computation has been discussed in the literature (Bains, 2003). However in this field only first theoretical results exist up to now (Bargiela and Pedrycz, 2008) and the fundamental methodological limitations must be accepted for the time being.

Nevertheless, considering optimization and data mining as an integral methodology affords significant synergies (Meisel and Mattfeld, 2010). On the one hand, optimization increases the efficiency of a large number of data mining operations. On the other hand, data mining leads to an increased effectiveness of many optimization approaches. In the following sections the synergies are desribed in detail and illustrated by examples from the literature.

4.2 Efficient Data Mining

Preprocessing and the core data mining step each represent a proper application domain of optimization. Both of the major data mining steps may be conducted according to the optimization process model of Sect. 4.1.3. Note that the (data-)system under consideration and the data generating process must be distinguished carefully.

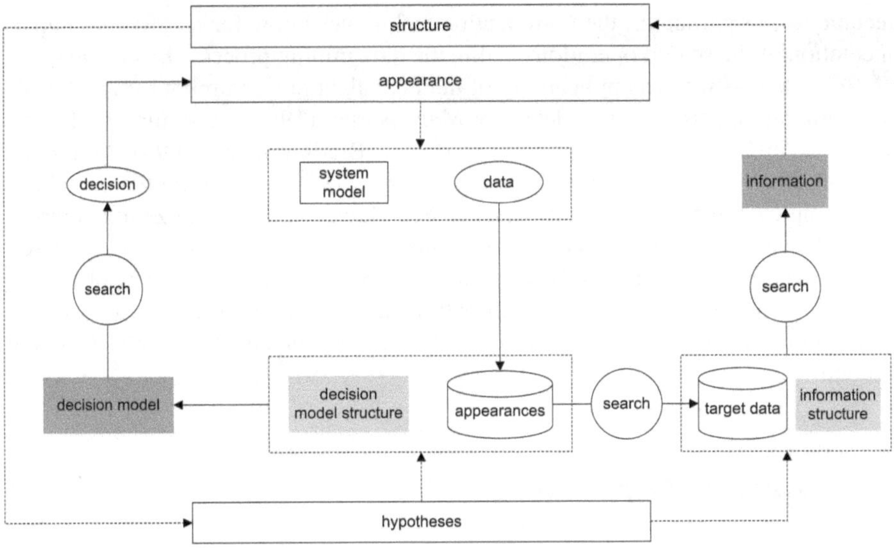

Fig. 4.2 Unified view of optimization and data mining. *Solid line arrows* indicate maps, partially implemented by search operators. *Dashed line arrows* represent deductions made by the user.

This distinction is particularly important if data mining is conducted in one go after a set of data has been collected from the generating process. In this case the (data-)system model identified in step one of optimization comprises a single appearance only, with the collected data representing constant attribute values. From an optimization perspective one single decision model is then derived as a basis of the search for a suitable target data set and the search for an instance of the information structure respectively.

A different (data-)system model is considered if data mining is not conducted in one go but incrementally. In this case the (data-)system model identified in step one of the optimization process comprises as variables the attributes defining the data generating process, i.e. an appearance of the data generating process is part of an appearance of the (data-)system under consideration. In contrast to the non-incremental case, a suitable target data set and accordingly an instance of the information structure is derived for each (data-)system appearance. To this end, the decision model of an appearance is often derived as an update of the decision model of the previous appearance.

Variations of the incremental case exist[1] in terms of the number of appearances of the data generating process recorded before an updated decision model is derived. In any case however, formulation of a decision model enables the use of optimization

[1] The field of incremental data mining attracted a lot of research interest especially throughout the past decade. An increasing number of new incremental techniques is developed for each of the data mining tasks. The use of these techniques is often denominated as "online data mining", "recursive data mining", "stream mining" or "real-time data mining". A survey is provided by, e.g., Stompe and Meisel (2006).

techniques. In particular, the optimization techniques allow for an efficient implementation of the search operations within the data mining process (Basu, 1998).

Relevance as well as applicability of mathematical programming for large-scale data mining are discussed in detail by Mangasarian (1997). A summary of early works about optimized data mining is due to Bradley et al (1999). Moreover, Padmanabhan and Tuzhilin (2003) give an overview of works within the domain of electronic customer relationship management. However, the optimization community discovered data mining as a major application domain only recently (Olafsson et al, 2008). Since then a rapidly increasing number of new works can be observed. The following sections illustrate both optimization for preprocessing (Sect. 4.2.1) and optimized information extraction (Sect. 4.2.2) by means of examples from the literature.

4.2.1 Optimized Preprocessing

Each of the preprocessing operations introduced in Sect. 4.1.2 has been considered from an optimization perspective in the literature. While only initial works exist for attribute *construction* (e.g., Piramuthu, 1996), a number of contributions fully implement either the selection operation, the projection operation or the transformation operation as an optimization procedure.

A *selection* operation serves the purpose of elimination of redundant data received from the data generating process. The identification of dissimilar data units is referred to as "de-duplication" in the data mining community. An example of de-duplication by optimization is given by Li and Jacob (2008). They define two attributes for evaluation of dissimilarity within the recorded data at hand. Then a bi-objective quadratic program is formulated as a decision model. The search operator for selection of an optimal target data set is implemented as a genetic algorithm.

The most prevalent *projection* operation is referred to as "feature subset selection". An example of efficient feature subset selection by optimization is due to Yang and Olafsson (2005, 2006, 2009). They introduce decision attributes for identification of a subset of the attributes originating from the data generating process to be included in the target data. A decision model is then derived in the form of a combinatorial optimization problem. The model is taylored to the use of the nested partitions heuristic (Shi and Olafsson, 2000) for searching the optimal attributes to be included into the target data set. The efficiency of this approach is shown by comparison to a standard operation for feature subset selection.

A different optimization approach to the feature subset selection problem is presented by Pendharkar (2006). His decision model leads to a constraint satisfaction problem. The attributes of the target data are selected in order to maximize the performance of a classifier subject to minimum evaluation costs. As a search operator a hybrid heuristic comprising simulated annealing and artificial neural networks is applied. In a related work (Meiri and Zahavi, 2006) the efficiency of feature subset selection is increased by application of simulated annealing as search operator.

Improvements are found in comparison to a standard projection approach relying on stepwise regression.

Attribute *transformation* is frequently conducted by mapping a continuous domain of an attribute into a discrete one. An example optimization approach to discretization is provided by Janssens et al (2006). They introduce decision attributes for determination of optimal boundary points for attribute values. The decision model is formulated as an integer program and corresponding optimization techniques are applied. An alternative decision model for discretization developed recently (Shen and Chen, 2008), relies on dynamic programming for the search of boundary points within a continuous attribute domain.

4.2.2 Optimized Information Extraction

Optimization approaches to the core data mining step of information extraction exist for every data mining task identified in Sect. 4.1.2. By far most of the works apply optimization to the tasks of descriptive modeling and predictive modeling. In contrast, both discovering patterns and rules and exploratory data analysis have been considered as optimization problems only recently. Hence only a few examples of optimization with respect to these tasks exist up to now. In the following, a number of examples that are both recent and representative is provided for each of the data mining tasks.

Descriptive Modeling

Most of the literature on optimization for descriptive modeling focuses on cluster analysis. Cluster analysis aggregates the appearances of the data generating process within the target data set by grouping them according to a similarity measure. The connection between cluster analysis and optimization by mathematical programming is known since long. Rao (1971) was one of the first to derive decision models for cluster analysis. Elements of the target data set are assigned to a group via decision attributes. The resulting similarity of elements within each group is evaluated according to a predefined similarity measure. The original decision models proposed by Rao imply the use of linear and nonlinear integer programming methods for cluster analysis. Most of the recent publications on cluster analysis by optimization focus on one of three aspects:

- *Enhanced search:* Sağlam et al (2006) formulate a mixed integer program for cluster analysis and propose a heuristic search procedure. Computational experiments on real-world data show this approach to outperform the well-known k-means algorithm (MacQueen, 1967) in terms of accuracy. Kulkarni and Fathi (2007) compare the properties of three different mixed integer decision models for cluster analysis. They derive conditions for optimal cluster analysis by use of a branch and cut technique as search method. Another notable approach with respect to search is presented by Hansen et al (2002). They derive a polynomial dynamic programming algorithm for application of cluster analysis to a data generating process with single attribute appearances.

- *New applications:* Innis (2006) focuses on cluster analysis taking into account the time order of the occuring data. One cluster must contain contiguous appearances of the data generating process. An integer programming decision model is derived and corresponding optimization techniques are selected as search operators. The approach is applied to cluster analysis of data generated by weather and aviation dynamics. Boginski et al (2006) show that clusters of similar stocks may be represented as cliques and quasi-cliques respectively in a financial market graph. They formulate an integer programming for clique partitioning and specify an heuristic optimization technique for cluster identification.
- *Similarity measures:* Beliakov and King (2006) derive new measures for evaluation of similarity. The measures are included into a bi-level programming decision model and a discrete gradient search method is applied for the conduction of cluster analysis. This optimization approach is shown to be capable of identifying non-convex overlapped d-dimensional clusters, a property present in only a handful of experimental methods before.
  In another work reflecting the benefits of optimization, Romanowski et al (2006) develop a similarity measure for bills of materials. The problem of calculating the similarity of a pair of bills is formulated as a binary integer program and the properties of an efficient heuristic optimization technique are discussed. Details about integration of this similarity measure into standard approaches to cluster analysis are given by Romanowski and Nagi (2004).

Predictive Modeling

One of the first works explicitely considering predictive modeling from a mere optimization perspective is due to Mangasarian (1965). Since then, a large body of contributions mainly focusing on mathematical programming emerged. The variety of approaches published recently ranges from goal programming for the purpose of two class classification (Jones et al, 2006) to nonlinear programs for multicriteria fuzzy classification (Belacel et al, 2007). Even general optimization frameworks for classification (Zhang et al, 2008) are developed. Two major aspects of recent optimization approaches to predictive modeling are exemplified in the following:

- *Enhancement of traditional methods:* The multi-class classification approach of Üney and Türkay (2006) is an example of the enhancement of traditional predictive modeling by optimization. Traditional classifiers separate the elements of the target data by hyperplanes. In contrast, the proposed mixed integer programming approach introduces the concept of hyperboxes for definition of class boundaries. Computational results show the optimization procedure to increase both accuracy and efficiency of multi-class data classification.
- *Original contributions:* Support vector machines (Burges, 1998) are an example of a predictive modeling technique originally relying on optimization. In its

basic form, a support vector machine comprises a decision model that may be represented as a quadratic program. Recently a number of enhancements of the original formulation of a support vector machine have been developed.
For instance, Trafalis and Gilbert (2006) formulate a robust support vector machine as a linear and a second order cone program. Subsequently an interior point method is implemented as search operation. In another contribution, Carrizosa and Martin-Barragan (2006) extend the theory of support vector machines by accounting for two types of misclassification costs. To this end, a biobjective quadratic program is established and a search method for determination of the set of pareto optimal solutions is provided.

Exploratory Data Analysis

Visualization of large amounts of high dimensional data is a major challenge in exploratory data analysis. It requires projection of data into a lower-dimensional space subject to conservation of a given measure of distance between different appearances of the data generating process.

The works of Abbiw-Jackson et al (2006) as well as of Bernataviciene et al (2006) are recent examples of visualization from an optimization perspective. The former transform the problem of visualization into a quadratic assignment problem and present a divide and conquer local search heuristic for efficient solution. As a precondition the approach requires discretization of attribute values in the preprocessing step. Bernatavičienė et al. realize an efficient visualization technique by combination of self organizing maps (Kohonen, 1995) and optimization based projection techniques. Different ways of combination are considered as well as evaluated according to a number of efficiency criteria.

Discovering Patterns and Rules

Not many contributions to discovering patterns and rules appeared in the optimization literature up to now. An example approach proposing an efficient search method is due to Baykasoğlu and Özbakir (2007). They develop an evolutionary programming technique for association rule mining. The technique is applied to five target datasets and shown to be very efficient and robust in comparison with standard association rule search methods.

4.3 Effective Optimization

The effectiveness of optimization critically depends on the decision model under consideration. The characteristics of a decision model determine the type of optimization technique that may be applied. A decision model allowing for use of an exact algorithm leads to a more effective decision than a model restricting the set of optimization techniques to heuristic search. The effectiveness of optimization vanishes if a decision model cannot be identified at all. The lack of a decision model may be caused in two ways. Either the available hypotheses about system structure do not permit setting up a decision model structure or the decision model structure at hand cannot be mapped into an effective decision model. In both cases optimization remains ineffective but a decision may be effected with the aid of data mining.

The examples given in Sect. 4.3.1 illustrate the use of data mining in case a decision model structure cannot be established at all. In such cases a collection of information about system structure is substituted for a decision model. Typically this information is represented as a set of rules possibly derived by data mining. The rules are derived as an abstraction of system appearances but at the same time they may be considered as an abstraction of the decision models (Agafonov et al, 2009) that would have been consulted if a decision model structure could have been established. Section 4.3.2 comprises illustrative examples of data mining supporting the purpose of optimization in the presence of a decision model structure that cannot be mapped into an effective decision model. In such a case data mining is integrated into the optimization process model of Sect. 4.1.3 for realization of an approximate decision model. In particular the approximation results from integration of information about system structure into the present decision model structure.

4.3.1 Decision Model Substitution

Substitution approaches in the literature tend to be specific to their field of application. Most of them involve the data mining task of discovering patterns and rules. Basically two kinds of approaches appear. Either rules are derived for directly determining the values of decision attributes or the rules provide information that must be augmented with additional hypotheses before decisions can be made. In the following, effective optimization by decision model substitution is illustrated by representative examples from the traditional optimization domains of manufacturing, warehousing, supply chain management, health care management, marketing and finance. Note that beyond these a large number of more unconventional domains exist - among them emerging disciplines like genomics and proteomics within the field of life sciences.

Manufacturing

Li and Olafsson (2005) create decision rules for single machine production scheduling. A decision tree technique is applied for identification of dispatching rules. Given two arbitrary jobs, the rules determine which job to be dispatched first. The target data requires preprocessing system appearances in terms of both attribute construction and projection operations. In another manufacturing application Shao et al (2006) derive rules for the decision on product configuration alternatives in bicycle production. An association rule technique is applied to sales transaction data. The generated rules allow for identification of the alternatives to be produced under consideration of attributes characterizing potential customers.

Huyet (2006) derives decision rules for operating a job shop composed of five workstations. In a first step, system structure is unveiled by application of a decision tree technique to a set of appearances of the scheduling system. The resulting information is then augmented with hypotheses leading to the desired decision rules. Raheja et al (2006) develop rules for condition-based maintenance of a production system. An association rule technique is applied to recorded appearances of the production system. The resulting rules indicate the need for system maintenance operations by identification of a number of failure modes of the production system.

Warehousing

Wu (2006) addresses the problem of deciding on a subset of items to be stored in a high productivity order completion zone. The number of items in the subset must not exceed a given threshold and the items must satisfy a given minimum percentage of the order occuring. The problem is transformed into the well-known data mining problem of finding frequent itemsets. A standard data mining technique is then used for identification of a suitable itemset. This approach is shown to be more effective than a popular heuristic optimization technique for the problem. The improved effectiveness is due to the fact that the rules derived do not only consider information about the popularities of single items but also take into account correlations. A similar problem is treated by Chen et al (2005). They decide on the compilation of order batches in a warehouse. An association rule technique is used to discover correlations between customer orders. Order batches are compiled according to these correlations with respect to the capacity constraints of storage/retrieval machines. The approach outperforms a number of known heuristic optimization techniques for the problem. It realizes both a lower number of batches and a reduction of the distance covered by storage/retrieval machines.

Supply Chain Management

Tseng et al (2006) derive rules for supplier selection from recorded appearances of a supply chain by application of a rough set procedure. The rules allow for supplier selection based on more attributes than ‘price’ only. Moreover, tedious supplier assessment processes are avoided. Lin (2009) also proposes supplier selection by data mining. An FP-growth algorithm (Han et al, 2004) is applied for identification of sets of parts purchased together frequently by a company. Primary and secondary suppliers are then selected according to their capability of satisfying the company’s demand patterns.

Sawicki and Zak (2008) determine maintenance rules for a fleet of vehicles. In a preprocessing step, the values of a set of attributes characterizing a vehicle are discretized. Then, a rough set procedure is applied. It returns the minimum number of attributes necessary for deriving a set of decision rules for the assessment of a vehicle. The inferred rules are used to build an automatic decision support system for fleet maintenance.

Health Care Management

Delesie and Croes (2000) conduct data visualization for supporting decision making in health care management. A measure of similarity of hospitals represented by insurance reimbursements for a set of medical procedures is established. Then the hospitals are displayed according to similarity. The resulting figures are interpreted in order to derive decision guidelines. In a more recent approach Glowacka et al (2009) consider the outpatient no-show problem as faced frequently by outpatient clinics. They develop sequencing schemes for patients in order to optimize a combination of performance measures used at the clinic. The rules for sequencing are obtained from association rule mining and shown to result in significant improvements over standard optimization approaches to the problem.

Marketing

Cooper and Giuffrida (2000) decide on the amounts of stock needed for various products in retail stores with respect to upcoming promotion events. First a market response model is consulted for determination of the number of units required per product in a retail store. However, the model is only capable of taking into account part of the attributes relevant to the decision. Attributes characterizing either particular retailers or geography are not incorporated. Thus, a rule induction technique is applied considering these attributes. The set of discovered rules is then used to adjust the decisions generated by the market response model.

Finance

Greco et al (2002) derive decision rules for company financing. They show the characteristics of the involved system attributes to suggest a dominance-based rough set approach to rule determination. The resulting set of rules for financing decisions is shown to be the most concise and non-redundant representation of the structural knowledge available from the recorded appearances given.

4.3.2 Decision Model Approximation

This section is dedicated to decision model approximation from an optimization perspective. But the integration of information about system structure into the structure of decision models is of interest also from a data mining perspective. In that context integration allows for an evaluation of information about system structure. Kleinberg et al (1998) were among the first to propose data mining results to be evaluated in terms of their utility in decision making. However, up to now only first theoretical steps have been made, indicating potential for further research.

More advanced results are available from an optimization perspective. In this context, decision model approximation is required whenever the present decision model structure does not lead to effective decision models. Ineffectiveness occurs if the decision model structure cannot be instantiated at all, or equally, if the derived decision models do not allow for subsequent use of effective optimization techniques. In both of these cases the decision model corresponding to an original problem $P = (X, f)$ is approximated by a model corresponding to a proxy $\tilde{P} = (\tilde{X}, \tilde{f})$. Decision model approximation appears as an additional task within step two of the optimization process model of Sect. 4.1.3. It may be realized with respect to both decision attributes and evaluation attributes. The following examples from literature illustrate both of these variants separately.

Approximation via Decision Attributes

Campbell et al (2001) develop a decision model for assigning a sequence of catalogs to each customer of a direct marketing company. The company's profit must be maximized by selection of the most profitable catalog sequence for each customer under consideration of cross-customer constraints. This optimization problem is represented as a linear program (derived from a corresponding decision model structure). However, this decision model comprises a decision attribute for each of the company's seven million customers. Finding an optimal decision turns out to be computationally infeasible. Thus an approximation of the original decision model is considered. The approximation implies assigning a sequence of catalogs to a (relatively small) number of customer types instead of considering each single customer individually. The resulting decision model is tractable but cannot be established without information about each customer's type. Thus data mining is conducted for identification of customer types. In particular homogeneous customer groups are identified by cluster analysis. Customers within the same group are considered as being of the same type and consequently customers of the same type are assigned the same sequence of catalogs. On the one hand the resulting overall assignment does not suffer from computational difficulties. On the other hand, the assignment clearly is an approximate solution as different customers of the same type might not be as homogeneous as suggested by the cluster analysis.

Brijs et al (2004) consider optimal product assortment for a retail store. The objective of optimization is gross margin maximization subject to a given amount of both handling costs and inventory costs. Initially a binary integer program is derived as a decision model. Each of the decision attributes represents a decision of whether or not to select a single product. Although this approach leads to a feasible decision, it suffers from the fact that cross-selling effects are not considered. Thus a new decision model is established by introduction of additional decision attributes. Each of these attributes refers to a set of products frequently purchased together and the value of an attribute indicates whether or not such a set should be selected. Identification of the sets considered in the decision model is done by application of a pattern discovery technique to recorded sales data.

The new decision model does not consider infrequent itemsets resulting from random purchase behavior. Consequently the model's objective function provides only an appoximation of the actual objective pursued. Application of a branch and bound procedure shows the advantages of the new model. Products rejected by the initial approach due to low individual profitability are now selected and significantly increase profitability.

Approximation via Evaluation Attributes

Ehmke et al (2009) approximate a decision model for a city logistics application. In particular they consider routing a vehicle in an urban road network comprising more than a hundred thousand road links. As travel times in the context of city logistics prove to be both time- and link-dependent (Ehmke and Meisel, 2008a), they must be derived from empirical data. Typically this is done by calculation of a number of sample means for each road link of the network. Given a road link, one sample mean is derived per hour of a day and per day of the week (Ehmke and Meisel, 2008b; Ehmke et al, 2010). Consequently, a decision model for determination of a route within the given road network comprises several millions of travel time values for evaluation of the total travel time of a route. The associated memory requirements are challenging. However, definition of an approximate decision model leads to a significant reduction. The model relies on approximate evaluation of the total travel time associated with a route. The time- and day-dependent travel times are approximated by use of only one base travel time value per link. These base values are then discounted with respect to hour and day. Discount factors are determined with respect to four different classes of similar links with the classes being derived by cluster analysis. Empirical tests indicate actual travel times resulting from routing decisions derived from the approximate model to be only slightly worse than those resulting from the detailed model.

Chen and Wu (2005) consider an order batching problem. Order batches must be compiled such that distances covered in a warehouse are minimized subject to a given set of orders and a given capacity of the retrieval device. A binary integer program is formulated as an initial decision model. Evaluation attributes reflect the total of distances resulting from decisions on the groupings of orders. Although this approach leads to a feasible decision, it suffers from the fact that correlations among orders are not considered. Thus a new decision model is established by introduction of additional evaluation attributes. The additional attributes represent the sum of correlations of orders within the batches. Thus the objective function does not match the actual objective of minimization of distances any longer. Instead the original objective is approximated by focusing on both distances and correlations. Correlation values are derived by means of an association rule technique. The decision model including both types of evaluation attributes is tested on a number of problem settings. Results show a reduced total of distance as well as a lower number of batches required in contrast to the initial approach.

Chapter 5
Approximate Anticipation

For the vast majority of real-world dynamic decision problems, the value function V required for solution of optimization problems of type

$$P_t = \left(D_t(s_t),\ c_t(s_t,d_t) + \mathbb{E}\left[V_{t'}(s_{t'})|s_t\right]\right)$$

cannot be derived because of the limitations exposed in Sect. 3.4. As a consequence, effective decisions cannot be made deliberately and lowering the degree of anticipation turns out to be the only remedy. However, a lower degree of anticipation trends to result in an increased focus on specific features of the dynamic decision problem at hand. Roughly speaking, the lower the degree of anticipation the less general an optimization approach will be. Only a very small number of methodical paradigms exist that generally represent a subset of lower degree anticipatory optimization. Moreover most of these paradigms so far remain without a significant impact on dynamic decision making.

A prominent example is represented by the classifier system paradigm originally proposed by Holland (1975). Modern versions of classifier systems (e.g., Butz, 2002) may basically be used to derive anticipatory decisions for dynamic decision making (Sigaud and Wilson, 2007). Such systems determine a set of state based decision rules by guiding an evolutionary algorithm as good as possible according to the principles introduced in Sects. 3.2 and 3.3. However, the research on classifier systems shows a strong bias towards theory and successful applications relevant to the real world are very rare (Sigaud and Wilson, 2007; Urbanowicz and Moore, 2009). Moreover, the existing applications (Bull, 2004) focus on other problem domains than dynamic decision making. Hence, the relevance of classifier systems for actual dynamic decision making turns out to be rather negligible up to now.[1]

However, the paradigm of approximate dynamic programming (ADP) recently attracted some attention with respect to dynamic decision making. On the one hand ADP has been claimed to provide effective solutions for a fairly broad class of

---

1 The academic test problems currently challenging classifier systems in the literature are usually characterized by less than a hundred states (Butz and Lanzi, 2009).

aggregation is reducing the domain of the value function. Two states $i, j \in S_t$ being mapped to the same element k of the target set of g_t, are assigned to the same value $\tilde{V}_t(k)$.

The actual values $\tilde{V}_t(g_t(s_t))$ may be determined by any procedure compatible with the actor-critic framework introduced in Sect. 3.2.4. Consider Procedure 7 as an example. Inclusion of preprocessing operations for state space aggregation requires making decisions according to

$$d_t^{\star,n,m} = \arg\max_{d \in D_t(s_t)} \left[c_t(s_t, d) + \mathbb{E}\left[\tilde{V}_{t'}^{\pi^{n-1}}(g_{t'}(s_{t'})) | s_t \right] \right] .$$

At the end of each iteration n, the value functions $\tilde{V}_t^{\pi^{n-1}}(g_t(s_t))$ determining the current policy are updated according to new estimates $\hat{V}_t^{n,M}(g_t(s_t))$ which are still calculated as a kind of moving average. However, updates are carried out as

$$\hat{V}_t^{n,m}(g_t(s_t)) = \hat{V}_t^{n,m-1}(g_t(s_t)) + \gamma \left[C_t(s_t) - \hat{V}_t^{n,m-1}(g_t(s_t)) \right] .$$

Note that despite aggregation the transition function F still operates on the original state space and the system trajectories considered are still of the same type as they are in the case of perfect anticipation. Both transition function and state trajectories would change, if the state transition probabilities p were also aggregated. Aggregating p according to the aggregated state space leads to an aggregate dynamic decision problem. This problem may be small enough for application of the methods of Chap. 3 and the obtained exact value function may be considered as an approximate value function for the original disaggregate problem.

Beyond that, a common variant of aggregation, keeping the original probabilities p, consists in approximating each of the true value functions $V_t(s_t)$ by means of a weighted mean of value functions $\tilde{V}_t^i(g_t^i(s_t))$ derived from different aggregation functions g_t^i. However, no matter what type of aggregation to be conducted, the quality of the anticipatory decisions derived from an approximate value function will heavily depend on the underlying hypothesis. The most important structural relations between the original attributes of the state variable and the true value function V must be preserved by the preprocessing operations selected for aggregation.

5.1.2 *Predictive Modeling*

A more general way of value function approximation is achieved by fully incorporating the data mining process into the methodology of Chap. 3. The data mining task of predictive modeling is of particular relevance. It consists in deriving approximate value functions $\tilde{V}_t(g(s_t))$ from an information structure $\tilde{V}_t(r_t, g_t(s_t))$ with r_t being a vector of parameters to be determined and with $g_t(s_t)$ being a vector of preprocessing operations. Note that the elements of r_t fully determine an approximate value function $\tilde{V}_t(g_t(s_t))$ as an instance of the structure $\tilde{V}_t(r_t, g_t(s_t))$. Thus, if

r_t is available, an approximate value function no longer needs to be represented by explicitly specifying a value estimate for every single state $s_t \in S_t$.

As with aggregation based approximation, hypotheses play a crucial role. They must ensure that the selected information structure captures the most important structural relations between state attributes and their values. According to the data mining process model outlined in Sect. 4.1.2, hypotheses may be incorporated in two ways. On the one hand the information structure itself can only be selected based on hypotheses. On the other hand, additional hypotheses should be consulted for preprocessing the state attribute values such that they match the selected information structure. In particular these hypotheses lead to a number of preprocessing operations, where an operation $g^i_\sigma(\sigma)$ represents the i-th component of $g_t(s_t)$. The prevalent preprocessing operation in the context of data mining for value function approximation is attribute construction. In the literature, the constructed attributes are often referred to as "features" and the functions representing the preprocessing operations leading to the feature values are known as "basis functions".[2] Each basis function represents a variable of the information structure selected.

In practice the approximate value functions are mostly derived by means of conventional regression techniques. However, first approaches to value function approximation making use of other techniques appeared recently (e.g., Deisenroth et al, 2009). In the case of conventional regression, an information structure must be selected from a broad bandwith of possible choices ranging from functional forms linear in the parameters to complex neural networks.

As in the case of mere state space aggregation, the actor-critic framework serves as a basis for determination of the approximate value functions $\tilde{V}_t(g_t(s_t))$. However, such a function must be derived as an instance of an information structure $\tilde{V}_t(r_t, g_t(s_t))$ now. Each update of $\tilde{V}_t^{\pi^n}$ at the end of an iteration n comprises a new data mining model representing the relation between the attributes of a state and the corresponding state value subject to policy π^{n-1}. Consequently a core data mining step must be carried out in each iteration. As a result, an actor-critic method generates a finite sequence $(r_{t,n})$ of parameter vectors with one element of the sequence determined in each iteration n. The approximate value functions $\tilde{V}_t(r_{t,N}, g_t(s_t))$ derived from the last element $r_{t,N}$ of the sequence is used for decision model approximation as specified in Problem 5.1. A detailed illustration of such actor-critic methods is provided in Sect. 5.3.

5.2 Stochastic Gradient Updates

Stochastic gradient methods are a suitable way of deriving a value function update if the information structure considered fits the form of a continously differentiable function. Such updates are particularly interesting because they may be reduced to stochastic approximation as introduced in Sect. 3.2.3.

---

[2] The term basis function is used even if the operations do not form the basis of a vector space.

A stochastic gradient method is a special type of iterative gradient method. The general form of an iterative gradient method for minimization of a continously differentiable function $h(x)$ can be stated as

$$x^{n+1} := x^n + \gamma^n q^n \; .$$

A variety of valid choices of the (positive) stepsize γ^n as well as the descent direction q^n exist. A prominent example with respect to the descent direction is illustrated in Sect. 5.2.1. Moreover, in this section the class of gradient methods resulting from this type of direction is applied to determination of an approximate value function $\tilde{V}_t^\pi$ subject to a given policy π. Section 5.2.2 provides details about stepsize selection for gradient methods within actor-critic methods.

5.2.1 Steepest Descent

Steepest descent requires selecting $q^n = -\nabla h(x^n)$ among the various flavors of possible descent directions.[3] A steepest descent gradient method

$$x^{n+1} := x^n - \gamma^n \nabla h(x^n)$$

may be applied for determination of an instance $\tilde{V}_t^\pi\left(r_t^*, g_t(s_t)\right)$ of an information structure $\tilde{V}_t\left(r_t, g_t(s_t)\right)$ subject to policy π. The parameters r_t^* emerge as a solution of the optimization problem $P = (\mathbb{R}, h(r_t))$ where

$$h(r_t) = \mathbb{E}\left[\frac{1}{2}\left(\tilde{V}_t(r_t, g_t(s_t)) - Z(s_t)\right)^2\right] ,$$

and for each $s_t \in S_t$, $Z(s_t)$ is a random variable providing sample observations of the value of s_t. As the gradient of $h(r_t)$ with respect to r_t is

$$\nabla h(r_t) = \mathbb{E}\left[\left(\tilde{V}_t\left(r_t, g_t(s_t)\right) - Z(s_t)\right) \nabla \tilde{V}_t\left(r_t, g_t(s_t)\right)\right] \tag{5.3}$$

the steepest descent gradient method for solution of P is given by

$$r_t^{n+1} := r_t^n - \gamma^n \mathbb{E}\left[\left(\tilde{V}_t\left(r_t^n, g_t(s_t)\right) - Z(s_t)\right) \nabla \tilde{V}_t\left(r_t^n, g_t(s_t)\right)\right] . \tag{5.4}$$

In the presence of a way of generating random samples $z_i(s_t)$ of $Z(s_t)$, an iteration of Procedure 5.4 can be carried out incrementally. Instead of the deterministic gradient, as specified in Eq. 5.3, stochastic gradients of type

$$\nabla h(r_t) = \left(\tilde{V}_t\left(r_t, g_t(s_t)\right) - z_i(s_t)\right) \nabla \tilde{V}_t\left(r_t, g_t(s_t)\right)$$

[3] See Bertsekas (1999) for a thorough discussion of alternative descent directions.

are consulted. The stochastic steepest descent gradient method for solution of P is then given by

$$r_t^{n+1} := r_t^n - \gamma^n \left[\tilde{V}_t\left(r_t^n, g_t(s_t)\right) - z_n(s_t) \right] \nabla \tilde{V}_t\left(r_t^n, g_t(s_t)\right) . \tag{5.5}$$

In analogy to Sect. 3.2.3 this method may be considered as a Robbins Monro procedure for solution of $f(r_t) = r_t$, where $f_t(r_t) = r_t - \nabla h(r_t)$. As with the Robbins Monro procedures of Sect. 3.2.3, variations result from different ways of generating the random samples $z_i(s_t)$. In particular the whole set of TD(λ) approaches may be transferred to the stochastic steepest descent methods.

5.2.2 Stepsize Rules

The behavior of the stochastic steepest descent method of Eq. 5.5 heavily depends on the stepsize values γ^n. The stepsize may be modified subsequent to each update of the parameter estimate. In the literature a guidance for the modification of γ^n is often referred to as a stepsize rule. One type of stepsize rule was introduced already in Sect. 3.2.2. For the purpose of evaluation of a given policy π, Procedure 6 requires updates of type

$$\hat{V}_t^{\pi,n}(s_t) := \hat{V}_t^{\pi,n-1}(s_t) + \frac{1}{m_{s_t}} \left[C_t^m(s_t) - \hat{V}_t^{\pi,n-1}(s_t) \right] . \tag{5.6}$$

Note that this update is equivalent to an iteration of a stochastic steepest descent method for determination of the true value $V_t^\pi(s_t)$ of a state s_t. The applied stepsize is $\frac{1}{m_{s_t}}$ with m_{s_t} indicating the number of updates carried out so far for the estimate of the value of s_t.

At least from a theoretical point of view the two prevalent criteria for the assessment of a stepsize rule are both probability of convergence and speed of convergence. Procedure 6 is guaranteed to converge almost surely to the true values $V_t^\pi(s_t)$ because the stepsize of Update 5.6 satisfies the following conditions as first identified by Blum (1954). Convergence occurs almost surely if a sequence of stepsizes γ^n, with $0 \leq \gamma^n \leq 1$ for each update n, ensures

$$\sum_{n=1}^{\infty} \gamma^n = \infty , \tag{5.7}$$

$$\sum_{n=1}^{\infty} (\gamma^n)^2 < \infty . \tag{5.8}$$

A variety of stepsize rules compliant with these conditions may be derived. Among them, the moving average rule $\gamma^n = \frac{1}{n}$ proves to be optimal as it leads to the minimum variance unbiased estimator of the true value $V_t^\pi(s_t)$ of a state s_t. However, this result cannot be transferred directly to the gradient method of Eq. 5.5. In contrast to Eq. 5.6, Eq. 5.5 comprises quantities of possibly different units. If the unit of a parameter r is different from the unit of the value of a state, the stepsize must additionally serve the purpose of standardization. In this case, the moving average stepsizes must be replaced by stepsizes $\gamma^n = \frac{\alpha}{n}$ with $\alpha > 0$.

A critical prerequisite of the conditions for convergence consists in the availability of unbiased sample observations $z_i(s_t)$ of the random variables $Z(s_t)$. As long as π is left unchanged and the state value estimates are initialized by 0, TD(1) provides such sample observations in the form of the accumulated contributions, i.e., $z_i = C_t^m(s_t)$. In contrast, a sample observation including an estimate of the value of a successor state is likely to be biased if the states are sampled asynchronously. As an example consider TD(0) which in case of an approximation relies on sample observations of type

$$z_i = c_t(s_t, d) + \mathbb{E}\left[\tilde{V}_{t'}^{\pi^{n-1}}(s_{t'})|s_t\right] .$$

The problem of bias is getting even worse as soon as stochastic gradient updates are integrated into an actor-critic method. In an actor-critic method the state values that must be estimated evolve over the iterations. Policies are changing frequently and the initial estimates of the state values are possibly far away from the true values of the policy in action. As a consequence the observations considered for updating the estimates are likely to arise from a highly transient (non stationary) data series.

It is well-known that mostly the moving average stepsize rule does not work properly if applied under these conditions. In particular the stepsize tends to drop to zero too quickly in the presence of non-stationary observations. Thus a number of alternative stepsize rules are proposed in the literature. Typically two types of rules are distinguished. Either a stepsize rule is deterministic with the stepsize being modified subject to the number of updates so far. Or a stepsize rule is stochastic with the stepsize being adapted continuously to the deviation of the current observation from the estimate of the previous iteration.

A popular example of a deterministic stepsize rule is gained by generalization of the moving average rule. The generalized harmonic stepsize rule is given by

$$\gamma^n = \frac{a}{a+n-1} . \tag{5.9}$$

Selecting a parameter value $a > 1$ leads to a sequence of stepsizes with each element of the sequence being larger than the corresponding element of the $\frac{1}{n}$ rule. Increasing a leads to a decrease of the rate at which the stepsize drops to zero. The generalized harmonic stepsize rule satisfies the convergence conditions independent of the value of a selected. Nevertheless the value of a must be selected carefully with respect to the application at hand. Mostly some parameter tuning is inevitable at this point.

An alternative approach consists in setting the stepsize to a constant value $\gamma^n = \gamma$. The constant stepsize rule does not satisfy the criteria for convergence and also requires some tuning effort. Nevertheless the rule is often applied especially if the number of parameters or state values to be estimated within an actor-critic method is large. In such a case selecting a single constant stepsize value may be preferred to carefully tuning the properties of a more complex stepsize rule for each of the gradient methods involved.

Many alternative deterministic stepsize rules have been developed. Common rules are for example the polynomial stepsize rule with $\gamma^n = \frac{1}{n^a}$, McClain's stepsize rule (McClain, 1974) or the search–then–converge stepsize rule (Darken and Moody, 1991). Each rule offers particular features for overcoming the problems encountered with the $\frac{1}{n}$ rule at the expense of additional parameters. While McClain's rule requires one parameter for introducing a tradeoff between the generalized harmonic and the constant stepsize rules, the search-then-converge rule comprises three parameters for imposing a delayed learning effect. Beyond the deterministic stepsize rules, a variety of stochastic rules appeared in the literature. One of the earliest rules depending on the observations instead of on the number of updates only is due to Kesten (1958).

A thorough discussion of a variety of stochastic as well as deterministic stepsize rules is provided by George and Powell (2006). In addition, these authors present a stochastic stepsize rule referred to as the optimal stepsize algorithm (OSA). OSA is capable of identifying stepsizes that minimize the estimation error in the case of non-stationary observations. However, practical application of OSA may require a considerable amount of experimentation. Not least OSA comprises an instance of McClain's rule whose parameters must be tuned among others.

5.3 The Generalized Actor-Critic Framework

In Sect. 3.2.4 the actor-critic framework is introduced as a family of methods for determination of the value function of an optimal policy $\pi^{\star}$. Within such an actor-critc method, the estimates $\hat{V}_t(s_t)$ are updated continuously by means of stochastic approximation and should finally converge to the true state values $V_t^{\pi^{\star}}(s_t)$.

A significant generalization of this framework occurs if the actor-critic principle is consulted for deriving approximate state values from an information structure $\tilde{V}_t(r_t, g_t(s_t))$. Two levels of generalization are illustrated in the following sections. Section 5.3.1 extends the degrees of freedom with respect to value function updates for the case of $\tilde{V}_t(r_t, g_t(s_t))$ being compatible with a regression model. Section 5.3.2 goes beyond regression models by transferring the actor-critic principle to general types of information structures.

5.3.1 Regression Models

Extending the actor-critic framework to approximate value functions is straightforward if the information structure allows for stochastic gradient updates as introduced in Sect. 5.2. Procedure 9 provides an example. It represents an actor-critic method for estimating the parameters of an approximate value function on the post-decision states. Note that the procedure makes use of TD(1) and may therefore be considered as a direct extension of Procedure 8.

Differing from Procedure 8, a policy π^n is now represented by the parameters $r_t^{n,M}$. Application of $r_t^{n,M}$ to an information structure $\tilde{V}_t^d(r, g_t(s^d))$ leads to an approximate value function $\tilde{V}_t^{d,\pi^n}(g_t(s^d))$. Beyond the dissimilar representations of a policy, two major differences arise between Procedures 8 and 9. In Procedure 9 the actor's decisions are based on an approximate value function instead of being based on estimates of the true state values. Correspondingly, the critic of Procedure 9 updates the parameters of the approximate value function instead of updating the estimates of the true state values. Note that the parameter update comprises a separate stepsize for each parameter set r_t. Each time a set r_t is updated, the corresponding stepsize γ_t must be determined according to one of the rules of Sect. 5.2.2.

Procedure 9: An actor-critic method with stochastic gradient updates of the parameters of an information structure.

$n \leftarrow 0 \quad \forall t \in T$: initialize r_t^0
repeat
  $n{+}{+} \quad m \leftarrow 0 \quad \forall t \in T : r_t^{n,0} \leftarrow r_t^{n-1,M}$
  repeat
    randomly select $s_0 \in S_0 \quad k \leftarrow 0 \quad m{+}{+}$
    for $(t \leftarrow 0; t < \mathcal{T}; t{+}{+})$ **do**
      if *(condition for next decision)* **then**
        $d_t^{\star,n,m} \leftarrow \arg\max_{d \in D_t(s_t)} \left[c_t(s_t, d) + \tilde{V}_t^d\left(r_t^{n,0}, g_t(s_t^d)\right)\right]$
        $\tau_k \leftarrow t \quad k{+}{+}$
      end
      $s_{t+1} \leftarrow F(s_t, d_{\tau_k}^{\star,n,m})$
    end
    $\forall t \in T : r_t^{n,m} \leftarrow r_t^{n,m-1}$
    for $(i \leftarrow k; i \geq 0; i{-}{-})$ **do**
      $t \leftarrow \tau_i \quad t' \leftarrow \tau_{i+1} \quad t^\dagger \leftarrow \tau_{i-1}$
      $C_t(s_t) \leftarrow c_t(s_t, d_t^{\star,n,m}) + C_{t'}(s_{t'})$
      $r_{t^\dagger}^{n,m} \leftarrow$
        $r_{t^\dagger}^{n,m-1} + \gamma_{t^\dagger}^{n,m-1}\left[C_t(s_t) - \tilde{V}_{t^\dagger}^d(r_{t^\dagger}^{n,m-1}, g_{t^\dagger}(s_{t^\dagger}^d))\right]\nabla\tilde{V}_{t^\dagger}^d(r_{t^\dagger}^{n,m-1}, g_{t^\dagger}(s_{t^\dagger}^d))$
    end
  until $(m = M)$
until $(n = N)$

However, updating the parameters of an approximate value function is not tied to stochastic gradient methods. A number of alternatives exist. As an example the parameters may be updated by direct solution of a least squares problem. An actor-critic method relying on this type of update is outlined as Procedure 10. The actor of this procedure is identical to the actor of Procedure 9. Moreover, Procedure 10 also resorts to TD(1)[4] and both of the procedures trigger an update after simulation of M

[4] Transferring both Procedure 9 and Procedure 10 from TD(1) to general TD(λ) is straightforward.

Procedure 10: An actor-critic method with least squares updates of the parameters of an information structure.

$n \leftarrow 0 \quad \forall t \in T$: initialize r_t^0
repeat
  $n{+}{+} \quad m \leftarrow 0 \quad \forall t \in T : r_t^{n,0} \leftarrow r_t^{n-1,M}, \mathscr{C}_t = \emptyset$
  repeat
    randomly select $s_0 \in S_0 \quad k \leftarrow 0 \quad m{+}{+}$
    for $(t \leftarrow 0; t < \mathscr{T}; t{+}{+})$ **do**
      if *(condition for next decision)* **then**
        $d_t^{\star,n,m} \leftarrow \arg\max_{d \in D_t(s_t)} \left[c_t(s_t,d) + \tilde{V}_t^d \left(r_t^{n,0}, g_t(s_t^d) \right) \right]$
        $\tau_k \leftarrow t \quad k{+}{+}$
      end
      $s_{t+1} \leftarrow F(s_t, d_{\tau_k}^{\star,n,m})$
    end
    for $(i \leftarrow k; i \geq 0; i{-}{-})$ **do**
      $t \leftarrow \tau_i \quad t' \leftarrow \tau_{i+1} \quad t^\dagger \leftarrow \tau_{i-1}$
      $C_t^{n,m}(s_t) \leftarrow c_t(s_t, d_t^{\star,n,m}) + C_{t'}^{n,m}(s_{t'})$
      $\mathscr{C}_t = \mathscr{C}_t \cup \{C_t^{n,m}(s_t)\}$
    end
  until $(m = M)$
  for $(t \leftarrow 0; t < \mathscr{T}; t{+}{+})$ **do**
    $r_t^{n,m} \leftarrow \arg\min_{r_t} \sum_{C_t \in \mathscr{C}_t} \left(C_t - \tilde{V}_t^d \left(r_t, g_t(s_t) \right) \right)^2$
  end
until $(n = N)$

trajectories. The only difference consists in how the parameter values are updated. Procedure 10 does not execute stochastic gradient updates. Instead it collects the sample observations occurring within the previous M trajectories. Each set $\mathscr{C}_t$ of sample observations assigned to time t is then fitted to an information structure. To this end a least squares problem must be solved for each of the sets. The new policy π^n is then represented by the parameter values resulting as the solutions of these problems in iteration n.

5.3.2 General Information Structures

Any approach that may be subsumed under the actor-critic framework serves the purpose of determination of approximate state values. However, provided that the information structure selected is powerful enough, the approximate state values may even converge to the true state values $V^{\pi^\star}$.

Throughout the past decades some research has been spent on the analysis of actor-critic methods relying on an information structure compatible with regression models. Thus, a number of settings are known to converge almost surely while error bounds have been derived for other settings. These results strongly underline the

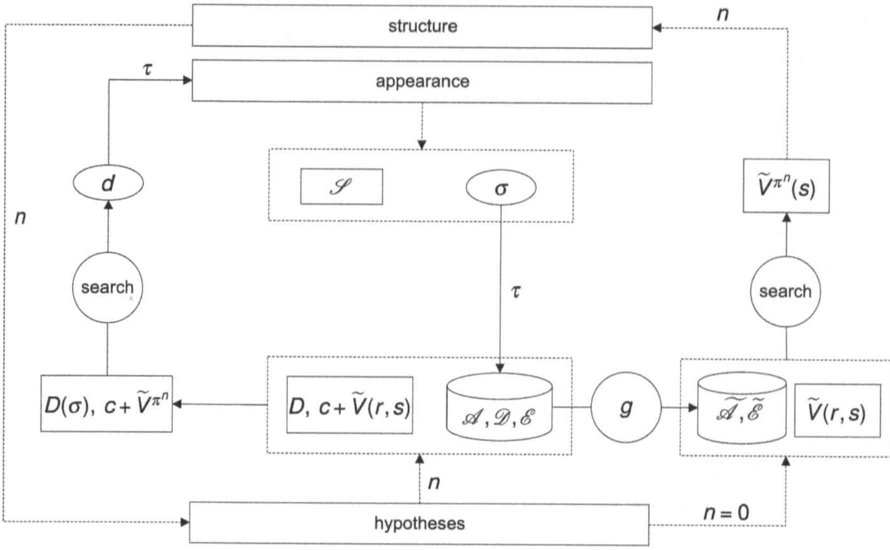

Fig. 5.1 The actor-critic framework in terms of optimization and data mining.

methodological soundness of the actor-critic framework. However, inspite of their theoretical value, such results are often of little help with respect to practical application. In particular, it is hardly possible to instantaneously derive an actor-critic approach to a particular application. The actor-critic framework turns out to be rather a guideline than a cookbook comprising predefined recipes. Powell (2007, Sect. 13.7) even notices that a successful actor-critic method for an important problem class is a patentable invention that represents a specific technological breakthrough.

Figure 5.1 highlights a number of challenges arising if an actor-critic approach shall be established for a particular application. Note that the figure represents a general actor-critic method in terms of the integral methodology introduced in Sect. 4.1.3.

Both an information structure and a set of preprocessing operations appear as elementary components. The preprocessing operations must match the information structure which is subject to hypotheses about system structure. Moreover, the information structure is incorporated into a decision model structure which serves as a template of appearance-specific decision models. Compilation of the entire decision model structure is an independent challenge comprising the blueprints of both the sets of feasible decisions and the objective functions of the decision models. Additional degrees of freedom occur with respect to the required search operators. On the one hand a search operator is needed for deriving a decision from a decision model and on the other hand a search operator is necessary for deriving an instance of the information structure.

On the basis of these elements an actor-critic method proceeds as follows. At a decision time τ the current values $a_i \in \mathscr{A}_i$ are acknowledged as part of the current system appearance σ. These values determine the decision model associated with

σ. The remaining elements of σ are then derived from a decision and its evaluation resulting from this model. System appearances occuring within an iteration n are recorded and preprocessed. The resulting target data contributes to determination of an updated approximate value function from the information structure. The value function represents a relation between a system state and its value subject to the policy currently applied by the actor. As soon as this structural insight is gained, it serves as a hypothesis for future decision making that is integrated into the decision model structure. The adapted decision model structure implies a change of the actor's policy.

The process model of Fig. 5.1 is independent of the type of information structure applied. So far the literature basically focuses on functional forms that are continuously differentiable. However, alternative actor-critic methods relying on more general types of information structures may be established.

Chapter 6
Dynamic Vehicle Routing

Vehicle routing is one of the major tasks in the context of transportation. The predominant influences currently inducing a change in transportation operations are outlined in Chap. 1. In particular these influences lead to an increased economic relevance of dynamic vehicle routing. A company whose business requires transportation operations may realize substantial cost reductions by considering vehicle routing problems (VRPs) as dynamic decision problems. The resulting problem class is important and elements of this class can be found in a large variety of economic sectors. Consider the routing of vehicles for the delivery of industrial products or for provision of field service for instance. Alternative examples comprise express courier services, the routing of vehicles within container terminals or taxicab services just to name a few.

Due to its broad relevance dynamic vehicle routing is considered for the experimental investigation of anticipatory optimization in the following chapters. The present chapter provides an introduction to dynamic vehicle routing. At first, the foundations of the field are illustrated in Sect. 6.1. Then an overview of the state of the art of dynamic vehicle routing is given in Sect. 6.2. To this end approaches from the literature are considered from the point of view of anticipation. Section 6.3 is dedicated to a particular type of dynamic vehicle routing problem. This problem serves as a case study for illustrating the impact of anticipatory decisions. Moreover, it provides the foundation of the following chapters.

6.1 Foundations

The class of dynamic vehicle routing problems (DVRPs) arose within the context of the more traditional approaches to vehicle routing. A brief outline of this background is given in Sect. 6.1.1. Subsequently, Sect. 6.1.2 provides a definition of dynamic vehicle routing and gives an impression of the broad variety of DVRPs that may occur in an economic context.

6.1.1 Vehicle Routing Background

The work of Dantzig and Ramser (1959) is widely considered as the first scientific treatment of vehicle routing. It deals with the routing of a fleet of gasoline delivery trucks between a bulk terminal and a large number of service stations supplied from the terminal. The truck routes to be determined should ensure that station demands are satisfied and the total milage covered by the fleet is minimal. Moreover the routes are established once before the first truck leaves the terminal. Both the shortest distances between any of the relevant geographical locations and the demands of the service stations are assumed to be fully known.

The gasoline truck problem is a generalization of the traveling salesman problem (TSP).[1] Thus, the true origins of vehicle routing may be attributed to the special case represented by the TSP. The TSP was formulated for the first time in a seminar talk at Princeton University in 1934 (Flood, 1956). It is concerned with the determination of the shortest roundtrip visiting each of a number of given geographical locations exactly once. Again the shortest distances between any of the locations are assumed to be known. Since the early days of vehicle routing a tremendous effort has been put in the formulation of VRPs as optimization problems as well as in the development of suitable optimization techniques. Different branches of research emerged.

One of the major branches focuses on more advanced vehicle routing problems.[2] The resulting optimization problems take into account an increased number of attributes of vehicles, drivers, customers and of infrastructure relevant to the routing task at hand. Nevertheless, each problem still represents a VRP[3] in the sense that the purpose still is determination of optimal routes for a number of vehicles. Advanced VRPs may for example be established by assigning a time window to each of the locations to be visited (e.g., Bräysy and Gendreau, 2005), by considering a vehicle for both pickup and delivery operations (e.g., Savelsbergh and Sol, 1995) or by making routing decisions subject to the drivers' working hours (e.g., Goel, 2009).

The compilation of Golden et al (2008) gives an impression of the recent progress in the formulation of advanced vehicle routing problems. At the same time it provides an overview of the recent developments with respect to optimization techniques suitable for solution of VRPs. As an optimal solution may typically be derived for only the smallest instances of VRPs, most of the research on optimization techniques for VRPs is about heuristics and metaheuristics. The development of improved optimization techniques for VRPs may be considered as an independent branch of research. As an example, a huge number of results (Applegate et al, 2007) have been derived for the TSP alone. In addition to heuristic techniques these results

---

[1] Dantzig and Ramser (1959) explicitly introduce the gasoline truck problem as a generalization of the TSP.

[2] Advanced vehicle routing problems are often referred to as "rich vehicle routing problems" in the literature (Hartl et al, 2006).

[3] In the following an explicit distinction between the (real) vehicle routing problem and its abstract representation will only be made if necessary. For the sake of simplicity the term "vehicle routing problem" refers to the optimization problem representing the real problem from now on.

comprise e.g. lower bounds and well-solvable special cases of the TSP (Burkard et al, 1998).

The TSP was also the starting point for the branch of research known as dynamic vehicle routing.[4] In a seminal paper about dynamic vehicle routing Psaraftis (1988) poses a number of dynamic traveling salesman problems (DTSPs). Just as the TSP, a DTSP consists of one vehicle and a number of geographical locations. Moreover, the shortest distances between any pair of locations are assumed to be known. However, in contrast to the TSP, the information about which of the locations need to be visited becomes known only gradually over time. Service requests may for example occur according to an individual Poisson process at each location. In any case, routing decisions must be made while the vehicle already is en route. A possible DTSP-objective may be serving as many requests as possible within a given period of time.

6.1.2 Dynamic Vehicle Routing Problems

Briefly speaking, any VRP becomes a DVRP as soon as it is considered as a dynamic decision process. Psaraftis (1988) expresses this by defining a VRP to be dynamic if information on the problem is revealed to the decision maker or is updated *concurrently* with the routing of the vehicles. Based on this definition he points out the difference between a dynamic TSP and a merely probabilistic traveling salesman problem (PTSP). A PTSP (Jaillet, 1985, 1988) is derived from a TSP by assigning an individual probability of request to each of the geographical locations. Only those locations actually requesting for service must be visited by the vehicle. The PTSP-objective consists in bringing the total set of locations into a sequence such that the expected total distance to be traveled for visiting only the requesting locations is minimized. Dropping the non-requesting locations from this sequence leads to the actual route of the vehicle. The PTSP is not dynamic because the information about the occuring requests is assumed to become revealed before the vehicle moves for the first time. Consequently the actual route of the vehicle is fully known from the beginning.

Initially the research on dynamic vehicle routing evolved quite slowly (Psaraftis, 1995). Part of the reason for that was the fact that the nature of DVRPs had not yet been fully understood. Dror et al (1989) are among the first to establish a link between a specific DVRP and a Markov decision process. However, the theory of Markov decision processes is consulted primarily for descriptive reasons. Neither suitable optimization techniques nor computational results are provided. In a more general paper, Powell et al (1995) integrate dynamic vehicle routing into the context of stochastic and dynamic decision models in logistics. Though a number of papers on the properties of DVRPs appeared (e.g., Dror, 1993; Séguin et al, 1997), the nature of the problem class has not been fully captured until the contribution of Powell

[4] In the literature dynamic vehicle routing is sometimes referred to as "online vehicle routing" or as "real-time vehicle routing".

et al (2001). They consider DVRPs as a subset of the much more general class of dynamic resource transformation problems (DRTPs). Moreover they provide a comprehensive modeling language capturing any essential aspects of a DRTP and being compatible with the theory of Markov decision processes.

Considering vehicle routing as a dynamic decision process establishes a significant generalization of the class of VRPs. This generalization enables companys to take into account customer behavior at a much more detailed level than before. As in the example of the DTSP, a DVRP may incorporate the case of previously unknown late customer requests. In addition, the case of initially unknown quantities demanded may be covered. Information about individual quantities may be revealed at the moment the corresponding request becomes known. Alternatively this information may appear at a later point in time. Moreover, a single customer may request for service either at most once or several times within the same period of time. In the latter case the demand of successive requests may either be accumulated or a new request may occur only after its predecessor has been served. Various scenarios may occur also with respect to the need to serve a customer request. In some cases service may be mandatory for each of the requests, while in other cases a request may be rejected or postponed. Obviously the number of dynamic vehicle routing problems matching different types of customer behavior is large. A comprehensive taxonomy allowing for classification of the (dynamic) vehicle routing problems considered so far in the literature is provided by Eksioglu et al (2009).

The various characteristics of a DVRP are usually combined with one of two types of objectives. In most of the cases a decision maker aims at directly maximizing his benefits, i.e., he wants to maximize the ratio of the demand served per time or per driven distance. In practice this objective is often pursued by maximizing the demand served within a predefined period of time or by minimizing the time for serving a fixed number of customers. Bertsimas and van Ryzin (1991) were among the first to propose an alternative objective. They discuss a number of dynamic vehicle routing scenarios with the service level provided to the customers being more important than direct maximization of benefits. In particular, they propose a variant of the DTSP known as the dynamic traveling repairman problem (DTRP). In the DTRP the routing of the vehicle must minimize the expected average amount of time elapsing between the occurence of a request and the arrival of the vehicle at the corresponding location. This objective is of particular interest for example in the case of emergency services.

A few steps have been made towards measuring the magnitude of the dynamics inherent in a DVRP (Kilby et al, 1998; Larsen et al, 2002, 2007). However, up to now no universal measure covering each possible DVRP has been developed. Larsen et al (2002, 2007) define a measure applicable to DVRPs comprising late requests and time windows. Based on that they propose a framework for classification of such DVRPs according to both magnitude of dynamics and objective pursued.

6.2 State of the Art

Throughout the past decade the research on dynamic vehicle routing has been picking up speed considerably. Today, a rapidly increasing number of publications can be observed. Contributions range from theoretical analyses (e.g., Ausiello et al, 2001; Archetti et al, 2003) to modules for large-scale decision support systems with an application background (e.g., Savelsbergh and Sol, 1998). Summaries of early works are due to Gendreau and Potvin (1998) as well as due to Bianchi (2000). A picture of the growing importance of the field results from contrasting these early summaries with the more recent survey papers of Ghiani et al (2003) and Larsen et al (2008). The latter differentiate explicitly between the initial phase of research on dynamic vehicle routing and the subsequent phase of extensive research starting in the year 2000.

Clearly, perfect anticipation is not an option for dynamic vehicle routing problems of practical relevance. Moreover for the vast majority of DVRPs even lower degrees of anticipation often turn out to be quite hard to realize. Nevertheless anticipation is the decisive factor for dynamic vehicle routing and the issue increasingly coins the DVRP literature. Nearly every author having contributed to the field proves to be aware of the significance of anticipation. The following sections provide a classification of approaches from the literature. This classification basically adheres to the prototypical degrees of anticipation introduced in Sect. 2.3.2. Omitting perfect anticipation and providing a more detailed distinction of approaches within the category of non-reactive anticipation leads to four categories:

- *Approximate anticipation* considers an approximation of the values of successor states as a criterion for making a decision at the current point in time.
- *Implicit anticipation* considers a (usually quite coarse) approximation of the values of successor states for identification of the set of feasible decisions at the current point in time.
- *Probabilistic non-reactive anticipation* explicitly considers both general hypotheses about the decision process and probabilistic information about upcoming events for making a decision. The values of the respective successor states are not considered specifically.
- *Conventional non-reactive anticipation* considers only non-probabilistic general hypotheses about the decision process for making a decision. Again, the values of the respective successor states are not considered specifically.

Of course a more detailed classification could be established by taking into account a larger number of intermediate degrees of anticipation. Yet, the four categories considered direct the attention to the most important characteristics of an approach with respect to anticipation.

The following sections present approaches from the literature in an ascending order with respect to their degree of anticipation. Note, however, that an approach realizing a higher degree of anticipation does not necessarily lead to better solutions with respect to the decision maker's overall objective. For example, if powerful general hypotheses about the decision process are available, a decision maker may be more successful by relying on a kind of non-reactive anticipation instead of introducing an approximation of the values of successor states. In some cases even a myopic approach optimizing the current contribution may be beneficial (Powell et al, 2000). However for most of the dynamic vehicle routing problems powerful general hypotheses are hardly available and the myopic approach fails. Moreover it is a fact that perfect anticipation never leads to solutions that are worse than solutions derived by any other approach. Thus, approximating perfect anticipation as good as possible is the natural task arising in the context of dynamic vehicle routing. The relevance of this task is underlined by many contributions from the literature. The studies of Thomas and White III (2004, 2007) illustrate the significance of a high degree of anticipation by demonstrating the superiority of perfect anticipation to non-reactive routing procedures commonly used in practice.

6.2.1 Conventional Non-reactive Anticipation

A quite early contribution to the dynamic vehicle routing literature is provided by Gendreau et al (1999). Being motivated from courier service applications, they consider the problem of routing a fleet of vehicles subject to dynamic customer requests with soft time windows. The time windows are soft in the sense that service is possible even if a vehicle arrives at a customer location later than permitted by the customer's time window. Nevertheless, a penalty occurs in case of a late arrival. Initially each of the vehicles is located at one single depot where it must return by the end of the depot's time window. The overall objective consists in minimizing the weighted sum of the total distance traveled and the penalty due to late arrivals. To this end every vehicle route is updated each time a new request becomes known. The updated routes are determined by a tabu search procedure that focuses on minimization of the contribution function. Thus, it does not anticipate the remaining decision process. However, anticipation is realized by the allocation of waiting time. If a vehicle's current route leads to an early arrival at the next customer location to be visited, the vehicle must wait at it's current location instead of moving on immediately. Thus the vehicle movement is performed at the latest possible time in order to allow for last minute changes of the planned route due to new arriving customer requests.

Ichoua et al (2000) show that the approach of Gendreau et al (1999) may be improved by allowing for diversion of vehicles, i.e., the current destination of a vehicle may be reset while the vehicle already is on its way. Moreover, a slightly different version of the problem is treated by Chen and Xu (2006). They strictly prohibit the violation of time windows and recalculate the routes of the vehicles at equidistant points in time. The applied solution technique relies on column generation instead of tabu search, but it does not introduce additional anticipatory features.

Angelelli et al (2009, 2010) consider a DVRP with two types of customer requests occuring during the course of a fixed number of days. A request either requires service at the day it appears or it allows for being postponed to the next day.

The overall objective consists in serving as many requests as possible at the expense of a preferably small total of distances traveled throughout the time horizon. Routes are updated at the beginning of every day as well as at fixed time intervals during a single day. An optimization problem for updating the routes anticipates the remaining decision process by considering one out of a number of proposed short term objectives. A short term objective is represented as a weighted sum combining various criteria for evaluation of either the routes of the present day or the routes of the next day. Example criteria are the distances traveled throughout a day as well as the number of currently known postponable requests served the next day. The approach requires calibration of a number of parameters for introducing a trade-off between the criteria of a short term objective. Experiments indicate that a short term objective evaluating the routes of both the present day and the next day is superior to more myopic short term objectives.

Branke et al (2005) elaborate on anticipation by waiting. They derive both theoretical results and a number of waiting strategies for dynamic vehicle routing subject to a single late requesting customer. A fleet of service vehicles is available and every vehicle must return to the depot at the same given point in time. The single late request is assumed to appear at a uniformly distributed random location within a convex service region. Under this assumption the authors prove the problem of finding an optimal waiting strategy to be NP-complete. Moreover, they apply seven different waiting strategies to a set of test instances and compare them by their relative frequencies of successful inclusion of the late request. The results highlight the value of a good waiting strategy and show that the empirically best performing waiting strategy may be derived from theoretical considerations on a simple two vehicle example.

An empirical study of four waiting strategies is presented by Mitrovic-Minic and Laporte (2004). They consider a dynamic pickup and delivery problem with time windows and dynamically occuring customer requests. The late requests must be included into the routes of a fleet of vehicles such that the overall sum of distances traveled for serving all requests is minimized. Both cheapest insert and tabu search are repeatedly applied for generating the vehicles' routes. Waiting time may be scheduled at any customer location as well as at the depot. As basic strategies the authors consider allocating the whole amount of waiting time either at the beginning or at the end of a route. In addition they propose two waiting strategies for realizing an intermediate balance between these two basic approaches. The computational results show the intermediate strategies to outperform the basic strategy of allocating the whole amount of waiting time at the end of a route.

In a follow-up work on the same DVRP, Mitrovic-Minic et al (2004) combine the use of waiting strategies with anticipatory evaluation criteria for vehicle routes. Both short term and long term objectives are considered as criteria for routing decisions. As a consequence a weighted sum of criteria occurs with a number of paramters to be determined empirically. Computational experiments show the new procedure to be advantageous especially in the presence of small problem instances. Another empirical study on the use of waiting strategies is provided by Kiechle et al (2008). They treat a dynamic pickup and delivery problem with time windows occuring in the context of patient transportation. Computational experiments indicate that adapting the waiting strategy to the current situation is superior to strictly allocating waiting time according to a predefined rule.

Waiting is also an option in the approach of Cheung et al (2008). They consider a dynamic pickup and delivery problem that differs from the previous ones with respect to the dynamic events considered. Both late customer requests and updates of the estimated travel time within the network may occur. Moreover, vehicle capacities must not be violated and the overall objective consists in minimization of travel time. However, the authors neither delve into waiting strategies nor do they establish anticipatory evaluation criteria. Instead their focus is on quick heuristic procedures for dynamic updates of the vehicle routes as new events occur. The resulting approach is found to be very effective for the problem instances considered although the realized degree of anticipation is quite low.

Bertsimas and van Ryzin (1991) and Papastavrou (1996) analyze dispatching policies for the dynamic traveling repairman problem. The analyses are based on queueing theory and result in lower bounds on the optimal expected system time of a customer request under the assumptions of either light or heavy traffic. Moreover, optimal dispatching policies are proposed for the case of light traffic. These policies anticipate the remaining decision process by frequently sending the vehicle to a waiting position located in the center of the service region. As a consequence the expected travel time for serving the next customer request is minimized if customer locations are uniformly distributed in the service region. Note that light traffic is a crucial assumption for the success of these policies. Bertsimas and van Ryzin (1993) extend their analysis to other versions of the DTRP. In particular, they consider the case of a fleet of vehicles as well as the case of a fleet of vehicles subject to capacity constraints. Beyond that, Swihart and Papastavrou (1999) provide a similar analysis with respect to different variants of a single vehicle dynamic pickup and delivery problem.

6.2.2 Probabilisitic Non-reactive Anticipation

Van Hemert and La Poutre (2004) consider dynamic routing of a fleet of vehicles over a given time horizon. The vehicles must pick up loads at customer locations and deliver them to a central depot within a given time window. Each vehicle has a limited capacity and may visit a customer location even in case no request occurred at this location. The overall goal is to deliver as many loads as possible to the depot. For this purpose the vehicles' routes are reoptimized each time a new customer request occurs as well as each time a vehicle executed either a pickup or a delivery operation. A route comprises two types of moves. On the one hand, the vehicle visits customers requesting for service while on the other hand it may also perform anticipated moves to customers that did not request for service yet. The inclusion of anticipated moves is based on knowledge about so called fruitful regions. A cluster of customer locations is considered as a fruitful region, if the locations together have a high potential for upcoming requests, where the potential is determined by the cluster's customer request probabilities. Anticipatory moves are included if a vehicle passes by a fruitful region while serving actual customer requests. Two evaluation criteria are combined into an objective for reoptimization. This objective enforces both a high number of the currently available requests to be served and a penalty for such anticipated moves that must be dropped from the route due to constraint violation. Computational experiments illustrate the potential of using knowledge about fruitful regions. In particular they show that the benefits gained from fruitful regions depend on the size of the time window for delivery.

A different way of exploiting knowledge about customer request probabilities is realized by Ichoua et al (2006). They treat the same DVRP as Gendreau et al (1999) and basically adopt the solution procedure of the latter. However, Ichoua et al (2006) provide a more thorough model of the problem and upgrade the solution procedure by inclusion of a threshold-based waiting strategy. As a basic prerequisite the service region is divided into a number of distinct zones. Moreover, each of these zones is assigned a probability indicating that at least one late request will occur within a certain time period. A vehicle that could wait at its current customer location is allowed to do so if the probability for a new request in the vehicle neighborhood within a certain period of time is greater or equal to a predefined threshold value. In this case waiting time is allocated subject to the latest feasible departure time as well as subject to a predefined upper bound for waiting time and a parameter representing some tolerance to additional lateness. The authors conduct experiments with several parameter settings and conclude that their procedure provides significant improvements over the conventional non-reactive procedure of Gendreau et al (1999).

A quite similar approach is due to Larsen et al (2004). They consider a DTSP with time windows and focus on the minimization of lateness. Every occuring customer request must be served and the vehicle route is recalculated each time a late request becomes known. Service is subject to soft time windows in the sense that upper bounds may be violated. However, the vehicle must wait if its route implies an early arrival at the next customer location. The available waiting time may be spent either at the current customer location or at one of a number of predefined idle points. If waiting is required an idle point must be selected and then the decision must be made whether to wait at this idle point or to spend the waiting time at the current customer location. Three rules for selection of an idle point are proposed. Moreover, the decision of whether or not to move to the idle point is based on the probability of receiving at least one new request close to the idle point within the available waiting time. The vehicle moves to the idle point if this probability exceeds a predefined threshold value. The authors conduct a number of computational experiments and conclude that their approach leads to a significant reduction of lateness at the extent of minimal increases of the distance traveled. Moreover, they give the overall recommendation to select the nearest idle point relative to the vehicle's current position and determine a relatively low threshold value.

Thomas (2007) proposes five waiting strategies for dynamic routing of a vehicle with late customer requests. Late requests must be either confirmed or rejected.

However, the decision on confirmation or rejection of a request may be postponed until the end of the given time horizon. If a confirmation is made, the corresponding customer is integrated into the current route of the vehicle according to a predefined sequence of all possible customer locations. The overall objective consists in maximization of the total number of requests served. Confirmations and rejections can be made each time the vehicle arrives at a customer location. At the same time a decision on whether to wait at the current position or move on to the next customer location must be made. Two of the waiting strategies considered for making this decision make use of customer request probabilities. In particular, those customers that may be served but did not request for service yet are aggregated into one single location, referred to as the center of gravity. A customer location's contribution to the center of gravity is weighted by it's request probability. The author illustrates the underlying decision problem by showing both a situation when the vehicle should always wait at the current location as well as a situation in which the vehicle should never wait. Moreover an optimal policy is identified for the one customer case. The empirical results comparing the different waiting strategies show that customer location information is more valuable than knowing the customers' request probabilities.

6.2.3 Implicit Anticipation

Only a small number of approaches to dynamic vehicle routing arrange the current set of feasible decisions according to an estimate of the value generated by the remaining decision process. An example is due to Bent and van Hentenryck (2004). They dynamically route a fleet of vehicles subject to late customer requests with time windows. The goal is to maximize the number of served customers, where the customer requests must be either confirmed or rejected after becomming known. At each point in time a set of alternative routing plans is maintained. One of these plans is executed for actually routing the fleet of vehicles. Each time a new event occurs, the set of plans is reoptimized and a new plan is selected for execution. Selection is done by means of a consensus function that identifies the plan being most similar to the alternative plans available. Thus the decisions to be made with respect to routing heavily depend on the whole set of routing plans that is currently maintained. Each time a new plan is added to the set of alternatives, a sample scenario of the remaining process is consulted. In particular the probability distributions of the customers that did not request for service yet are sampled from the current point in time to the end of the time horizon. Subsequently the routing plan is derived as a projection of a plan comprising both the requests currently available and the future requests suggested by the scenario. Clearly a single sample scenario provides only a weak estimate of the value generated by the remaining decision process. Nevertheless the authors find the sample scenario approach to produce dramatic improvements over non-reactive approaches. Bent and Van Hentenryck (2007) show that more effective versions of the sample scenario approach can be generated by integration of waiting and relocation strategies. Moreover, a slightly different sample scenario approach is presented by Hvattum et al (2006). They consider a real-world case in the context of a norvegian distribution company leading to a DVRP similar to that treated by Bent and van Hentenryck (2004). An improved version of the approach is discussed by Hvattum and Lokketangen (2007).

More recently, Ghiani et al (2009) consider a sample scenario approach to a dynamic pickup and delivery problem. They aim at minimizing the expected overall customer inconvenience in terms of customer system time. Any time a new customer request occurs, a number of alternative routing plans is constructed by insertion of the request at different positions of the current routing plan. The resulting plans are then evaluated with respect to the system time of their constituting customer requests. Those routing plans offering the lowest system times are selected and reconsidered for evaluation by means of sample scenarios. In particular a number of sample scenarios of the near-future demand are generated and integrated into each of the preselected routing plans. Each preselected plan is then evaluated by the sample average of the system time of both the known requests and the near-future requests generated by the correpsonding scenario. According to this criterion, the best routing plan is selected for implementation. Anticipation is realized by considering the sample averages which provide rough estimates of the value generated by the remaining decision process. However, the main criterion for selection of a routing plan still is the system time of only the known customer requests. Computational experiments show the sample scenario approach to clearly outperform an approach without anticipation. Moreover, the authors show that determining the number of sample scenarios by means of a fully sequential indifference zone selection procedure (Kim and Nelson, 2001) leads to a considerable reduction of the computational burden.

6.2.4 Approximate Anticipation

Approximate anticipation for dynamic vehicle routing is at the very beginning. Up to now, approaches that fully realize approximate anticipation are very rare. Nevertheless a number of problems related to vehicle routing have been treated recently. Maxwell et al (2010) provide an approximate policy iteration approach to an ambulance redeployment problem. In particular, they address the question of where to redeploy idle ambulances in order to minimize the number of unserved emergency calls over time. Topaloglu and Powell (2006) realize approximate anticipation for effective repositioning of business jets to serve uncertain future customer requests. Their approach consists in approximating the value function by means of a hybrid of linear and piecewise-linear functions. While the preceding works focus on applications that are quite large-scale, Bertazzi et al (2001) treat a problem concerning a smaller number of entities. They propose approximate policy iteration for a stochastic inventory transportation problem, where one vehicle must carry out the transportation of products from a supplier to a customer subject to inventory costs.

A related problem with slightly more entities is considered by Proper and Tadepalli (2006).

Truckload trucking is quite closely related to vehicle routing. Considering truckload trucking as a dynamic decision process means that at any point in time there is a set of available truck drivers as well as a set of loads that must be moved from one city to another. Loads appear dynamically over time and a decision maker must assign one driver to one load subject to minimal costs. Basically costs result from moving empty to a pickup location as well as from actual transportation of loads from their origin to their destination. Spivey and Powell (2004) realize approximate anticipation for the problem by representing it as a dynamic assignment problem and proposing an approximate value iteration procedure for solution. A remarkable real-world case application of this approach is provided in a follow-up paper by Simao et al (2009).

A realization of approximate anticipation for a dynamic vehicle routing problem is provided by Secomandi (2000). In contrast to the contributions referred to in the previous paragraphs, he considers a DVRP where all the customer locations to be visited are known in advance. Late requests do not appear. Instead, the quantity demanded by a requesting customer becomes known only as soon as the vehicle arrives at the customer's location. All the customers are served by a single vehicle with limited capacity. To this purpose the vehicle is located at a single depot at the beginning and must return to the depot both in case its capacity is exceeded and after having served the last open customer request. The vehicle routing determines both the next customer location to visit and whether the vehicle passes by the depot on its way to this location. The routing aims at minimization of the expected distance traveled for satisfaction of every request.[5] Secomandi (2000) proposes an actor-critic method with a linear information structure for value function approximation. Computational results show this approach to outperform a non-anticipatory benchmark procedure. Additionally a computationally efficient heuristic technique for generating state value estimates is introduced. Application of these estimates for anticipatory optimization produces better results than the actor-critic method. Extended variants of this approach are provided by Secomandi (2001) and Novoa and Storer (2009). Moreover, Secomandi (2003) consideres a similar approach for a traveling salesman problem with stochastic travel times. Recently Secomandi and Margot (2009) reconsidered the problem treated by Secomandi (2000). The proposed approach is based on state space aggregation and outperforms the method applied by Secomandi (2001).

Approximate anticipation for dynamic vehicle routing with late customer requests was not considered until recently. The first, and up to this writing only, approach is provided by Meisel et al (2010, 2011). They realize approximate anticipation for dynamic routing of a single vehicle subject to late requests and service confirmations. Details about the work of Meisel et al (2010, 2011) are not presented

[5] The problem considered by Secomandi (2000) is sometimes contrasted with dynamic vehicle routing problems. However, this distinction seems questionable as the term "dynamic vehicle routing" is independent of the type of information being revealed over time. Rather it makes a statement about how routing decisions are made.

at this point as their approach is covered by the following Chap. 7. A discussion of the vehicle routing problem they consider is provided in the following section.

6.3 Dynamic Routing of a Service Vehicle

The present section considers a dynamic vehicle routing problem with one vehicle and a number of known customer locations. The vehicle is operated during the course of a given period of time. At the end of this period the vehicle must return to a dedicated depot while at the beginning it is located at a start depot. Customers are divided into the two distinct sets of early requesting customers and late requesting customers. Early requesting customers are known to request for service at the beginning of the time period, whereas late requests appear randomly over time according to known individual request probabilities. Late requests must be either confirmed or rejected directly after becoming known. Both early requests and confirmed late requests must be served before the vehicle returns to the depot. Part of the late requesting customers may not request for service at all. The overall objective is maximization of the expected number of customers served.

The number of applications resulting in a DVRP of this type is large. Consider the provision of on-site field service or the delivery of commodities for instance. Moreover, a specific pickup application directly results from the less–than–truckload example of Sect. 1.2. In general, the problem may be considered as dynamic routing of a service vehicle subject to late requests. A formal representation of this problem is needed before a solution may be derived. Thus, Sect. 6.3.1 establishes the corresponding Markov decision process. Based on this representation, the impact of anticipatory decisions is analyzed in Sect. 6.3.2.

6.3.1 Problem Formulation

Routing a service vehicle may be considered as a dynamic decision process. A formal representation of such a process is established by definition of the elements of a Markov decision process. However, as a precondition the system of objects that are involved into the decision process must be identified.

This system of objects includes a set $I = \{1,2,\ldots\}$ of potential customers with known geographic locations as well as a service vehicle. Moreover, system objects are given by both the start depot and the end depot as well as by the set of road links connecting any pair of customer locations. Depots and road links are assumed to consist of constant attributes only. The former are fully characterized by their locations, while the latter are each fully characterized by origin, destination as well as the travel distance between the two. For the sake of simplicity one unit of travel distance is assumed to correspond to one unit of time and the set $I_0 = \{0,1,2,\ldots\}$ comprises both the customers $\{1,2,\ldots\}$ and the start depot 0. The geographical

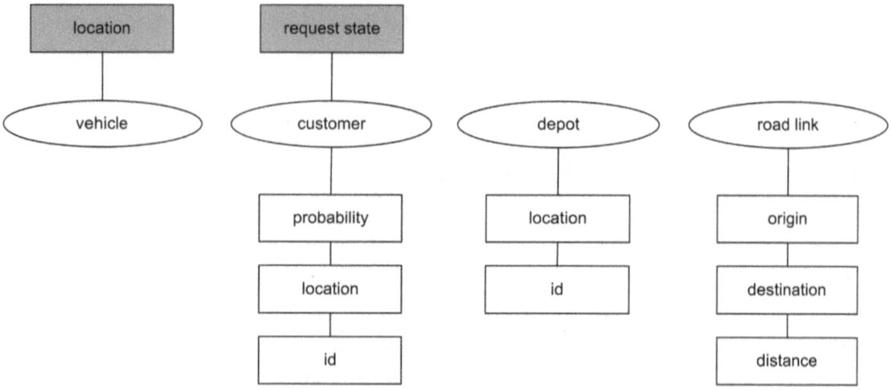

Fig. 6.1 System objects and their attributes. Attributes with a constant value are illustrated by *white boxes* while variable attributes appear as *gray-shaded boxes*.

location of a customer $i \in I$ is denoted as g_i, while the locations of the start depot and the end depot are denoted as g_S and g_E respectively.

The vehicle is routed over the course of $\mathscr{T} = |T|$ time units and its position at an arbitrary point in time t is denoted as $n_t \in I_0 \cup \{E\}$. At $t = 0$ the vehicle is located at the start depot and at $t = \mathscr{T}$ it must be located at the end depot, i.e., $n_0 = 0$ and $n_{\mathscr{T}} = E$. Moreover, the set I is divided into two complementary parts I_E and I_L, that is $I = I_E \cup I_L$ and $I_E \cap I_L = \emptyset$. I_E is the set of early requesting customers, while I_L represents the set of late requesting customers. The requests of early requesting customers are known at time $t = 0$. Each late requesting customer $i \in I_L$ is assigned a value α_i characterizing his request behavior. More precisely, α_i represents the probability of the occurence of a request of i within $\mathscr{T}$ time units. A customer requests for service at most once with the probability of request being equal for each time unit passing by.

An overview of the system objects and their attributes is provided in Fig. 6.1. In the following paragraphs this system serves as a basis for the definition of a Markov decision process for dynamic routing of a service vehicle.

State Space Definition

The state of the system comprises the current values of those system attributes whose values may vary over time. Thus, both the current vehicle position and the request states of the customers are considered. At each point in time the request state r_{ti} of customer i is characterized by one of the following values:

- $r_{ti} = 0$ if i has not requested for service yet
- $r_{ti} = 1$ if i has launched a request but has not been confirmed or rejected yet
- $r_{ti} = 2$ if the request of i has been confirmed
- $r_{ti} = 3$ if the request of i has been either rejected or served

At time $t = 0$, $\forall i \in I_E : r_{ti} = 2$ and $\forall i \in I_L : r_{ti} = 0$. By additionally consulting the current position n_t of the vehicle, the overall state $s_t \in S_t$ of the system at time t is represented as $s_t = (n_t, r_{t1}, r_{t2}, \ldots, r_{t|I|})$. As a consequence, the total set S of states, comprises $(\mathcal{T}+1) \times (|I|+2) \times 4^{|I|}$ elements. Thus, routing a service vehicle in the course of 100 time units with only 10 potential customers already leads to more than a billion states.[6]

Decision Space Definition

Adding a decision maker to the basic system of Fig. 6.1 requires definition of the options available at a decision time. In particular, the decision maker must decide on both the movement of the vehicle and the confirmation of customer requests.

A new decision on where to move the vehicle to must be made at those points in time at which the vehicle is located at either a customer location or at the start depot. In other words, if the vehicle is on its way to a specific location, it may not be diverted to a different place. Diversion is not incorporated because it is not an option in most of the real-world applications. The reasons for not considering diversion in a business context are manifold. For example, if the vehicle operates within the center of a city or within any other area that is subject to heavy traffic, it may be quite bothersome and even dangerous for the driver if he must continuously pay attention to new routing information shown by an electronic device. Particular difficulties may arise in the case of streets that do not allow for quick changes of direction as well as in the case of the service vehicle being a bicycle or a motorbike. During a bike ride an electronic device showing new routing information may not be easily accessible. Moreover, there may be at least short-term announcements of service to customers that are visited next. Such announcements would possibly be revised if diversion was allowed and a customer being subject to multiple revisions would probably consider the company's quality of service to be rather low.

Service quality also plays an important role with respect to decisions about the confirmation of customer requests. Three common scenarios may occur if a request is not confirmed but rejected instead.

- The rejected customer may be forced to address his request to another company.
- The rejected request may be postponed to the next day or to the next work shift, where it appears as an early request at $t = 0$.
- If the company operates a fleet of vehicles and the vehicles are assigned to close-by areas, a rejected request may be shifted to another vehicle. However, even if a larger fleet of vehicles is available, it is unlikely that any request can be served within the given period of time. Thus rejections in the sense of the first two scenarios may still occur.

In any case a requesting customer usually wants to know as fast as possible if his request will be served today, the next day or if he even has to look for another company

[6] Of course part of the states may occur only in theory.

to take over the job. Thus, from the point of view of service quality a confirmation or rejection should be made immediately after the request occurs. However, from an operational point of view, a company might prefer to make such a decision as late as possible because late decisions increase the company's operational flexibility. Moreover, as a direct consequence of prohibiting diversion, it is in general least beneficial for a company to accept or reject a customer while the vehicle is moving from one location to another. A reasonable tradeoff between a high level of customer service and operational flexibility might be realized by making a decision on confirmation of a request at the next point in time the vehicle is located at either a customer location or at the start depot. On the one hand this way of decision making maintains some operational flexibility because the company must not react immediately to a request. On the other hand a requesting customer may immediately be told a point in time in the near future, at which he will be informed about confirmation or rejection of his request.

In summary, a decision d_t at a point in time t consists of both a confirmation (respectively rejection) operation d_{tc} and a vehicle move operation d_{tm}, i.e., $d_t = (d_{tc}, d_{tm})$. Such a decision is made for the first time at $\tau_0 = 0$. Moreover, a decision is due each time the vehicle arrives at a customer location as well as each time it waited for one time unit at its current position. New customer requests occurring between two subsequent decision times τ_k and τ_{k+1} are treated as if they became known at τ_{k+1}.

At a decision time τ_k, each $i \in R_{\tau_k} = \{i | r_{\tau_k i} = 1\}$ must be either confirmed or rejected. Let $\mathscr{P}(R_{\tau_k})$ be the powerset of new requesting customers becoming known at τ_k and let $\bar{R}_{\tau_k} \subset \mathscr{P}(R_{\tau_k})$ be the sets whose elements can be confirmed without forcing a violation of the time horizon. Then the operation $d_{\tau_k c}$ may be represented as a set of customers receiving a confirmation at τ_k, i.e., $d_{\tau_k c} \in \bar{R}_{\tau_k}$. The set $d_{\tau_k c}$ comprises those customers making a transition from $r = 1$ to $r = 2$, and implicitly defines the customers making a transition from $r = 1$ to $r = 3$, which are those in $R_{\tau_k} \backslash d_{\tau_k c}$.

The operation $d_{\tau_k m}$ is represented as a single element $i \in I_0 \cup E$. The location of i indicates the place where the vehicle is expected to be at the next decision time τ_{k+1}. Either waiting at the current location or moving on to a customer with $r_{\tau_k i} = 2$ is possible. Thus, $d_{\tau_k m} \in M_{\tau_k} = n_{\tau_k} \cup \{i | r^d_{\tau_k i} = 2\}$. Note that M_{τ_k} depends on $d_{\tau_k c}$, and let the set $\bar{M}_{\tau_k} \subseteq M_{\tau_k}$ be the set of those move operations not causing a violation of the time horizon.

The set D_{τ_k} of feasible decisions at τ_k may be specified as

$$D_{\tau_k}(s_{\tau_k}) = \{d_{\tau_k} | d_{\tau_k c} \in \bar{R}_{\tau_k} \wedge d_{\tau_k m} \in \bar{M}_{\tau_k}(d_{\tau_k c})\} \,. \tag{6.1}$$

Optimality Equations

Every company aims at using its resources efficiently. Consequently, in many cases the primary objective of running a service vehicle will be serving as many customers as possible within the given period of time. Subject to this overall objective, the contribution received from a single decision d_t at a decision time t is given by the number of confirmed requests. This allows for formulation of the contribution functions c_t as

$$c_t(s_t, d_t, s_{t'}) = c_t(d_t) = |d_{tc}| \,. \tag{6.2}$$

As the customer behavior represents an exogenous influence that cannot be controlled by the decision maker, the overall goal results from Eq. 2.4. Accordingly, the decision maker aims at maximization of the expected sum of contributions

$$\bar{C} = \max_{D^{|T|}} \mathbb{E}\Big[\sum_{t \in T} |d_{tc}|\Big] \,. \tag{6.3}$$

Clearly, a state transition in the context of routing a service vehicle is determined by both the current decision d_t and the customer behavior. The function p representing the impact of the customer behavior on a state transition may be derived from the customers' request probabilitites α_i. Based on that, Bellman's equations may be consulted for stating the value of a state s_t as

$$V_t(s_t) = \max_{d_t \in D_t} \Big[|d_{tc}| + \mathbb{E}\big[V_{t'}(s_{t'}) | s_t\big] \Big] \,. \tag{6.4}$$

However, a customer's decision to issue a request is independent of the decisions d_t. Consequently, the two influences on a state transition can be considered as separate and Bellman's equations may be formulated around the post-decision state variables (see Fig. 3.2 in Sect. 3.3.2). A perfect anticipatory decision for dynamic routing of a service vehicle is given by

$$d_t^{\star} = \max_{d_t \in D_t} \big[|d_{tc}| + V_{t'}^d(s_{t'}^d) \big] \,. \tag{6.5}$$

Note that considering the post-decision state variables leads to a significant reduction of the size of the state space. Immediately after a decision is made, there are no requesting customers that did not get a confirmation or a rejection. Consequently the customer request state value $r_{ti} = 1$ may be dropped if post-decision state variables are consulted. The size of the state space reduces down to $(\mathscr{T} + 1) \times (|I| + 2) \times 3^{|I|}$ states.

6.3.2 Case Study

The following case study demonstrates the complexity of anticipatory decision making for dynamic routing of a service vehicle. A small instance of the problem introduced in the previous section is illustrated in Fig. 6.2. This instance comprises a total of five geographical locations. They resolve into the depot locations g_S and g_E as well as the locations of a set $I = \{1, 2, 3\}$ of potential customers. Each element of I represents a late requesting customer, i.e., $I = I_L$ and $I_E = \emptyset$. Moreover $\mathscr{T} = 25$ and every pair of geographical locations is assumed to be connected by a road link.

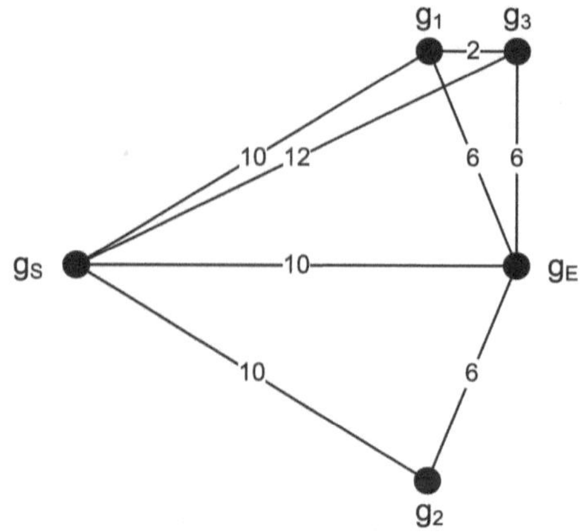

Fig. 6.2 Small instance of the problem of routing a service vehicle.

Figure 6.2 shows the road links together with the distances implied. The distances are assumed to be driving distances that must not necessarily be Euclidean. Note that the link connecting g_1 and g_2 as well as the link connecting g_3 and g_2 is omitted. Each of the two links is assumed to have a distance of 12, which means that their use is prohibitive with respect to the given time horizon.[7]

Obviously, this problem instance is quite small in comparison with most of the instances resulting from a practical background. Nevertheless an analysis of the small instance provides a good impression of the complexity of real-world applications. The latter may comprise dozens or hundreds of customers, which makes even only a rough analysis difficult to carry out. In contrast, the small instance offers a number of insights at first glance.

Clearly at most two customers may be served within the given time horizon. If either customer 1 or customer 3 already received a confirmation, customer 2 must be rejected in any case. On the other hand, both 1 and 3 must be rejected if 2 has already been confirmed. Moreover, if the first request at all occurs later than $t = 9$, any request must be rejected and the vehicle returns to the end depot without serving a single customer. If the first request at all is issued by either 1 or 3 such that the request can be served without violating the time horizon, it is always optimal to confirm this request.

However, in spite of these insights, identification of the optimal policy is not trivial even in the small instance. A full decision comprises the four basic options of confirmation of new requests, rejection of new requests, waiting at the current position or moving on to another location. The following example scenarios show that each of the four options may be beneficial. Every scenario assumes that the vehicle is located at g_S and that none of the customers has issued a request up to the current point in time.

[7] Recall that the vehicle is located at the start depot at $t = 0$.

- *Confirm.* If customer 2 issues a request at a point in time t_2 and the expected number of servable requests emerging from customers 1 and 3 later than t_2 is less than 1, then customer 2 should be confirmed. The vehicle may move on to g_2 immediately because additional confirmations are impossible without a violation of the time horizon.
- *Reject.* If customer 2 issues a request at t_2 and the expected number of servable requests emerging from customers 1 and 3 later than t_2 is more than 1, then customer 2 should be rejected. The vehicle must wait at g_S taking into account the possibility that no customer request will be served until the end of the time horizon.
- *Move early.* If customer 1 issues a request for example at $t_1 = 6$, it clearly must be confirmed. In addition the vehicle could either wait at g_S or move on to g_1 immediately. The waiting option would allow the vehicle to stay at g_S until $t = 9$. At $t = 9$ it would have to move on to g_1 and to the end depot without the possibility of visiting any other location. Customer 3 could be confirmed only if he issued a request at $t = 7$. On the other hand, if the vehicle moves on to g_1 immediately at $t = 6$ it will arrive at g_1 at $t = 16$. In this case, customer 3 can be confirmed if he issues a request at any point in time before $t = 17$.
- *Wait.* If customer 3 issues a request at $t_3 = 6$, it must definitely be confirmed. In addition the vehicle could either wait at g_S or move on to g_3 immediately. Moving on immediately would result in the vehicle arriving at g_3 at $t = 18$. No more requests occuring later than t_3 could be served. However, a request of customer 1 occuring at $t_1 = 7$ can be confirmed and served if the vehicle waits at g_S for one unit of time right after confirmation of customer 3. In this case, the vehicle can move on from the start depot at $t = 7$ and visit g_1 as well as g_3 before arriving at the end depot at $t = 25$.

In particular, an optimal policy for the small instance comprises the following reasoning. If every customer issues a request at the same time t_{123} with $t_{123} \leq 7$, then customer 2 must be rejected and both customer 1 and 3 are served. In case $8 \leq t_{123} \leq 9$, both customer 2 and 3 are rejected and only customer 1 receives a confirmation. Similar statements can be made for the case of customers 1 and 3 requesting for service at the same point in time t_{13}, up to which no request of customer 2 was received. Moreover, if the first requests are issued concurrently by customers 1 and 2 at $t_{12} \leq 9$, then customer 2 must be rejected and the vehicle should move to g_1 immediately. It then waits at g_1 as long as possible without causing a violation of the time horizon and serves customer 3, if he issues a request at time t_3 with $t_3 \leq 17$.

Concurrent requests of customers 2 and 3 lead to confirmation of customer 3 and rejection of customer 2, if the time t_{23} of request satisfies either $1 \leq t_{23} \leq 5$ or $t_{23} = 7$. In the former case the vehicle can move on to g_3 at once and wait there until $t = 17$ in order to serve a possible request of customer 1. As a possible request of customer 1 cannot be served in the latter case, the vehicle moves to the end depot via g_3 without waiting anywhere. A parallel request of customers 2 and 3 at $t_{23} = 6$ also leads to a rejection of customer 2. However, the vehicle must not move until $t = 7$, allowing for confirmation of a possible request of customer 1 occuring at $t_1 = 7$. Finally, a parallel request of customers 2 and 3 at $t_{23} \in [8,9]$ requires the rejection of customer 3 and the confirmation of customer 2, as only the latter may be served without a violation of the time horizon.

If the first request is issued at time $t_1 \in [8,9]$ by customer 1, this customer must be served without the possibility of accepting additional requests. On the contrary, an additional request of customer 3, occuring no later than $t = 17$ may also be served if the preceding request of customer 1 appeared at $t_1 \leq 7$. In case of the first request being issued by customer 3 at $t_3 \leq 5$, the vehicle should move to g_3 immediately as well as additionally serve a possible request of customer 1 occuring not later than $t = 17$. The case of customer 3 requesting for service at $t_3 = 6$ is discussed already within the preceding *Wait*-scenario. Moreover, if $t_3 = 7$, customer 3 is confirmed and served without accepting further requests.

Table 6.1 Properties of the small instance subject to the customers' request probabilitites.

α_1	α_2	α_3	t_2^*	$\bar{C}$
0.50	0.90	0.50	1	0.79486
0.80	0.90	0.90	2	1.20630
0.90	0.90	0.90	3	1.32906
0.95	0.90	0.90	4	1.42604
0.95	0.90	0.99	5	1.64924

Both the *Confirm*-scenario and the *Reject*-scenario treat the case of the first request being issued by customer 2. The optimal policy with respect to this case is heavily influenced by the customers' request probabilities. The request probabilities define the point in time t_2^* before which it is reasonable to reject customer 2 in anticipation of upcoming requests of customers 1 and 3. Table 6.1 specifies t_2^* for some combinations of request probabilities. In addition the table provides the expected number of confirmations $\bar{C}$ associated with each combination of request probabilities if an optimal policy is applied. Note that the value of t_2^* merely depends on the probabilities α_1 and α_3. On the contrary, the expected number of additional confirmations due to taking into account t_2^* is also heavily influenced by α_2. The more likely it is that the first request will be issued by customer 2, the bigger will be the impact of being aware of t_2^*.

Chapter 7
Anticipatory Routing of a Service Vehicle

The following sections introduce a selection of approaches to anticipatory optimization for dynamic routing of a service vehicle. As proposed in Sects. 2.3.2 and 6.2 the approaches are categorized according to their degree of anticipation. Section 7.1 covers perfect anticipation, which – from a practical point of view – is limited to small problem instances. Nevertheless, the realization of perfect anticipation for such small instances provides valuable insights with respect to approaches featuring lower degrees of anticipation.

A number of new approaches featuring a lower degree of anticipation are introduced in Sects. 7.2 and 7.3. In particular, Sect. 7.2 develops actor-critic methods for the problem of dynamic routing of a service vehicle. The resulting procedures match the concept introduced in Chap. 5 and realize approximate anticipation. Subsequently, Sect. 7.3 proposes a variety of non-reactive anticipatory approaches to dynamic routing of a service vehicle. Part of the non-reactive approaches are inspired by ideas present in the literature while others are entirely new. Note that the procedures of Sect. 7.2 represent the first approaches ever to realization of approximate anticipation for dynamic vehicle routing with late customer requests.

7.1 Perfect Anticipation

Perfect anticipation is realized by means of the concepts introduced in Chap. 3. As an example, the problem formulation presented in Sect. 6.3.1 basically allows for direct application of Procedure 8 in order to determine the state values needed for perfect anticipatory routing of a service vehicle. However, a successful realization of perfect anticipation requires some additional reasoning that is outlined in Sect. 7.1.1. Based on that, Sect. 7.1.2 provides empirical results showing that forward dynamic programming generates a perfect solution for the exemplary instance introduced in Sect. 6.3.1. Finally, Sect. 7.1.3 summarizes the limited effectivness of perfect anticipation with respect to larger instances of the problem of dynamic routing of a service vehicle.

7.1.1 State Sampling

At first glance applying Procedure 8 seems to be straightforward. However, a sufficient number of updates for every state must be guaranteed in order to permit identification of an optimal policy. Convergence to the true state values even requires that every state may potentially be visited infinitely often. Following the real-time dynamic programming principle as introduced in Sect. 3.2.1, Procedure 8 selects the next state to receive an update by means of making a decision of type

$$d_t^\star \leftarrow \arg\max_{d \in D_t(s_t)} \left[c_t(s_t, d) + \bar{V}_t^{d,\pi^{n-1}}(s_t^d) \right] . \tag{7.1}$$

Such a decision fully determines the next post-decision state, which will receive an update as soon as a new sample estimate of the value of the following pre-decision state is available. This means that the next state to receive an update heavily depends on the current value function estimates $\bar{V}_t^{d,\pi^{n-1}}(s_t^d)$ resulting from the previous iterations. However, exploiting the experience made up to the current iteration for chosing the next state to receive an update may turn out to be a pitfall. As an example, consider the following situation with respect to the small instance introduced in Sect. 6.3.2:

Assume that an adequate stepsize has been selected and Procedure 8 is applied for the case of $t_2^* = 3$. In this case, an optimal policy $\pi^\star$ must for example reject customer 2 if he requests for service at $t_2 = 1$ without any other customer having requested before. Thus, starting from state $s_1 = (0,0,1,0)$, $\pi^\star$ must chose state $s_1^d = (0,0,3,0)$ over both state $s_1^d = (0,0,2,0)$ and state $s_1^d = (2,0,3,0)$. This, however, at least requires $\bar{V}_1^{d,\pi^{n-1}}(0,0,3,0) > 1$, because otherwise the contribution c_1 received from confirmation of customer 2 cannot be exceeded.[1]

As the estimated values of the states $(0,0,3,0), (0,0,2,0), (2,0,3,0) \in S_1^d$ are initialized as $\bar{V}_1^{d,\pi^0}(0,0,3,0) = \bar{V}_1^{d,\pi^0}(0,0,2,0) = \bar{V}_1^{d,\pi^0}(2,0,3,0) = 0$, $s_1^d = (0,0,3,0)$ needs to be updated in order to permit $\bar{V}_1^{d,\pi^0}(0,0,3,0) > 1$. Unfortunately, such an update will never happen as long as the next state to be updated is selected according to a decision represented by Formula 7.1. However, updates can be enabled by means of a small modification of Procedure 8. Basically this modification consists in separating more clearly the issue of selecting the next state to receive an update from the issue of generating a sample estimate. An illustration of the resulting type of state sampling is given in Fig. 7.1. The figure shows one (*gray-shaded*) state trajectory, which is generated for the purpose of selecting states that receive an update. On the contrary, the non gray-shaded states within the figure are visited only for

[1] More precisely, the optimal policy requires $\bar{V}_1^{d,\pi^{n-1}}(0,0,3,0) > 1 + \bar{V}_1^{d,\pi^{n-1}}(0,0,2,0)$ as well as $\bar{V}_1^{d,\pi^{n-1}}(0,0,3,0) > 1 + \bar{V}_1^{d,\pi^{n-1}}(2,0,3,0)$. Yet, as no more confirmations can occur if the system is either in state $s_1^d = (0,0,2,0)$ or in state $s_1^d = (2,0,3,0)$, both $\bar{V}_1^{d,\pi^{n-1}}(0,0,3,0)$ and $\bar{V}_1^{d,\pi^{n-1}}(2,0,3,0)$ will always be equal to zero.

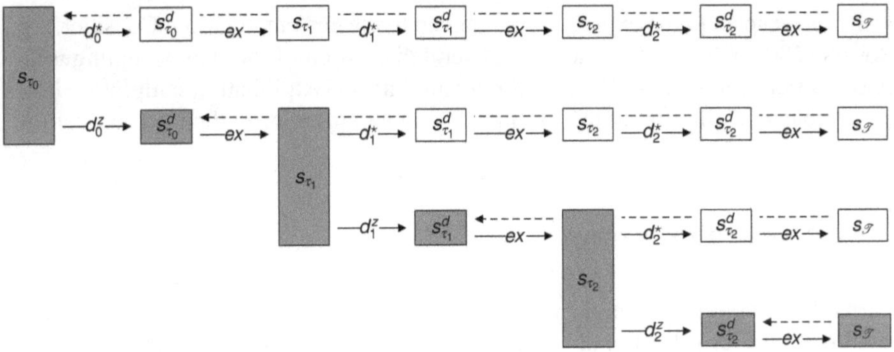

Fig. 7.1 Exploration sampling for TD(1).

the purpose of obtaining a sample estimate of the value of one of the gray-shaded post-decision states.

Starting from the initial state $s_{\tau_0} = (0,0,0,0)$ at the first decision time $\tau_0 = 0$ two decisions are made. The first decision $d_0^\star$ results from application of Formula 7.1 and represents the first step of a partial state trajectory which is generated for obtaining a sample estimate of the value of the initial state. The subsequent decisions required for reaching the terminal state $s_{\mathcal{T}}$ of this partial system trajectory are again made according to Formula 7.1. The single contributions occuring in the course of this trajectroy are used for updating the value of the pre-decision state s_{τ_0}, which at the same time may be considered as a virtual post-decision state, preceding s_{τ_0}. The second decision made in state s_{τ_0} is of type

$$d_t^z \leftarrow Z(D_t(s_t)) \,, \tag{7.2}$$

with $Z(\cdot)$ being a function that randomly returns one of the decisions within the current set of feasible decisions. That is, in state s_{τ_0}, $d_0^z \leftarrow Z(D_0(s_0))$. Assigning the same probability of selection to each of the feasible decisions at the current point in time allows for every possible successor post-decision state to appear. The state $s_{\tau_0}^d$, resulting from decision d_0^z, is selected as the next state to receive an update. In order to obtain a sample estimate for carrying out this update, another partial state trajectory is generated. This trajectory starts from state s_{τ_1} of the gray-shaded trajectory and evolves by making decisions according to Formula 7.1. Moreover, at the same decision time τ_1 a decision $d_1^z \leftarrow Z(D_t(s_t))$ is made in order to step forward to the next post-decision state to reveive an update. This alternating procedure of generating a post-decision state to be updated and a partial trajectory for obtaining a sample estimate is repeated until the terminal state of the gray-shaded trajectory is reached. Subsequently the procedure starts over from the initial state.

The key issue of this sampling procedure consists in the fact that the states to be updated are no longer selected by exploiting the experience gained from the previous trajectories. Instead the randomized decisions introduce a type of exploration sampling that enforces updates of even such states that do not seem to be worth a visit with respect to the experience gained up to the current iteration. Consider, e.g., Powell (2007, Chap. 10) for a more general discussion of the state sampling issue. An empirical comparison of exploration sampling and exploitation sampling follows in the course of solution of the exemplary problem instance in the next subsection.

7.1.2 Solution Properties

Perfectly solving the exemplary problem instance of Sect. 6.3.2 by means of Procedure 8 requires selection of an adequate stepsize rule. In order to try to keep the amount of time spent on stepsize selection at a low level, it makes sense to check out the performance of a simple constant stepsize first. Thus, in preparation of the empirical results illustrated in this section, a hundred iterations of Procedure 8 were run for each of five different constant stepsize values. As a result of these experimental runs a constant stepsize of 0.0005 was selected.[2]

Both Procedure 8 and its exploration sampling variant were applied with this stepsize. Moreover, a variety of settings of the customers' request probabilities were considered. Each of these settings shows similar effects, so that within the remainder of this section, the focus may be placed on one illustrative case. A nice illustration of the behavior of the two variants of Procedure 8 can be obtained for instance from the following setting. Let the individual request probabilities of both customer 1 and 3 be identical. In particular, assume that their probabilities of issuing a request within one single time unit are set to 0.125, which implies $t_2^* = 4$. Moreover, let the probability of a request of customer 2 within one unit of time be 4.5×0.125. This fairly large value is helpful for clearly demonstrating the impact of being aware of t_2^*.[3] Subject to this setting of request probabilities, an optimal policy must yield an expected number of $\bar{C} \approx 1.5687$ confirmations.

Figure 7.2 outlines the evolution of the expected number of confirmations over the first 1,000 iterations of both the exploitation variant and the exploration variant of Procedure 8. Both variants update the states of one single trajectory within one iteration. At each iteration, the expected number of confirmations of the current value function estimates $\forall t\ \forall s_t^d \in S_t^d : \bar{V}_t^{d,\pi^n}(s_t^d)$ is estimated by applying the corresponding policy π^n to the same set of 10,000 randomly generated test trajectories. Thus, in Fig. 7.2 a data point denotes the average number of confirmations over these

---

[2] The five stepsize values considered were 0.5, 0.05, 0.005, 0.0005 as well as 0.00005. The first value (0.5) represents an initial guess. After 0.0005 proved to be suitable and lead to the results presented in this section, the "neighboring" stepsize values of 0.0004 and 0.0006 were also considered, in order to get an idea of the sensitivity of Procedure 8 with respect to the stepsize value. Both 0.0004 and 0.0006 lead to similar overall results as 0.0005. Nevertheless, a full sensitivity analysis has not been carried out.

[3] Keep in mind, that the specific values of the request probabilities are primarily selected for illustrative purposes. The idea here is to present the solution of a case that is both quite challenging as a problem and suitable for getting an impression of the properties of a perfect anticipatory solution. However, the general behavior of the two variants of Procedure 8 is the same for any setting of request probabilities.

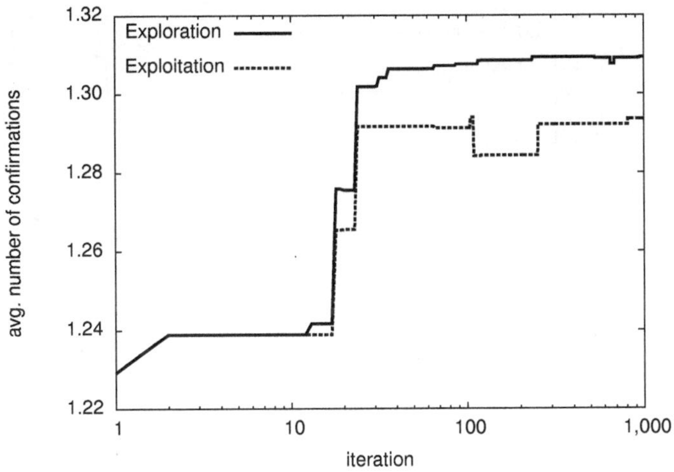

Fig. 7.2 Evolution of the solution quality over the first 1,000 iterations of Procedure 8 executing either exploration or exploitation state sampling.

10,000 trajectories. Up to iteration twelve, exploitation and exploration result into the same average number of confirmations. Subsequently, exploration continuously generates policies that are superior to their respective exploitation counterpart at the same iteration. Apparently exploration entails more changes of the solution quality than exploitation. Beyond iteration 25 the former shows many relatively small improvements, while the latter visits a smaller number of quality levels without clearly offering a trend.

A picture of the overall behavior of the two approaches is shown in Fig. 7.3. This figure outlines the evolution of the expected number of confirmations over 200,000 iterations. Again, policies are tested by the same 10,000 trajectories and the expected numbers of confirmations are estimated by averages. However, in contrast to Fig. 7.2, this time the quality of only every hundredth policy appears as a data point. Obviously the courses of the resulting lines are very different and exploitation performs worse than exploration. The quality of the exploitation state value estimates climbs up to 1.2947 until iteration 1,100 and remains nearly constant from there on. Only four rather small improvements occur beyond iteration 1,100, the last of which takes place at iteration 93,000 and leads to an overall quality of an average of 1.2952 confirmations. On the contrary, the exploration variant of Procedure 8 yields a policy of quality 1.3093 already at iteration 1,100 and stays at approximately this level until iteration 18,000. Subsequently the exploration curve is characterized by two jumps improving the quality up to an average of 1.5623 confirmations. It then stays at this level for quite a large number of iterations before it starts oscillating between 1.5623 and 1.5723.

A detailed analysis of the state value estimates occuring from iteration 93,000 on proves that the peaks of the exploration curve correspond to an optimal policy each. This analytical conclusion may additionally be cross-checked by generating a

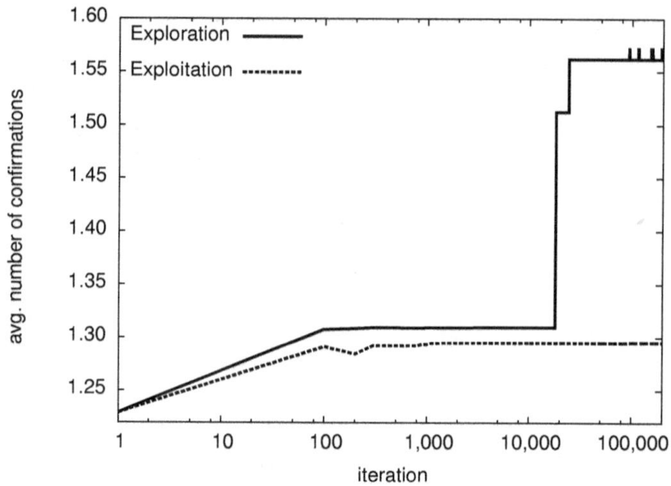

Fig. 7.3 Evolution of the solution quality over 200,000 iterations of Procedure 8 executing either exploration or exploitation state sampling.

more precise estimate of the expected number of confirmations resulting from the peak policies. Estimating the quality of these policies by means of a million test trajectories yields an average of 1.5681 confirmations. In view of the true value of an optimal policy being 1.5687, this empirical result can be considered as an acknowledgement of the conclusions gained from the preceding analysis of value estimates.

In summary, the exploration variant of TD(1) gets to an optimal policy after 93,000 iterations, while within a total of 100,000 iterations its exploitation counterpart does not even exceed an estimated level of 1.3 expected confirmations. The exploration variant does not stabilize exactly at the optimal policy. Instead, it temporarily drops back to a policy of inferior quality after having maintained the optimal policy for some hundered iterations. An explanation of this behavior may be derived from Fig. 7.4.

This figure illustrates the evolution of the value estimates for state $s_t^d = (0,0,3,0)$ at times $t = 1,\ldots,4$ over the first 100,000 iterations of the exploration variant of TD(1). As discussed in Sect. 7.1.1, an optimal policy requires $\forall t \in T | t < t_2^* : \hat{V}_t(s_t) > 1$. For $t = 1$ this condition is satisfied from iteration 18,100 on. Moreover, $\hat{V}_2(s_2)$ clearly overshoots 1.0 at iteration 24,100. Note that in Fig. 7.3 both iteration 18,100 and iteration 24,100 yields a jump of the estimated solution quality. In contrast to the value estimates of s_1 and s_2, the estimate of s_3 needs a significantly larger number of iterations until it exceeds 1.0. $\hat{V}_3(s_3) > 1$ is satisfied for the first time at iteration 93,000, which is exactly the same iteration at which the first peak can be observed in Fig. 7.3. However, $\hat{V}_3(s_3)$ drops back below 1.0 at iteration 93,300 before rebounding above 1.0 at iteration 93,600. Obviously a peak of the exploration curve of Fig. 7.3 occurs whenever $\hat{V}_3(s_3) > 1$.

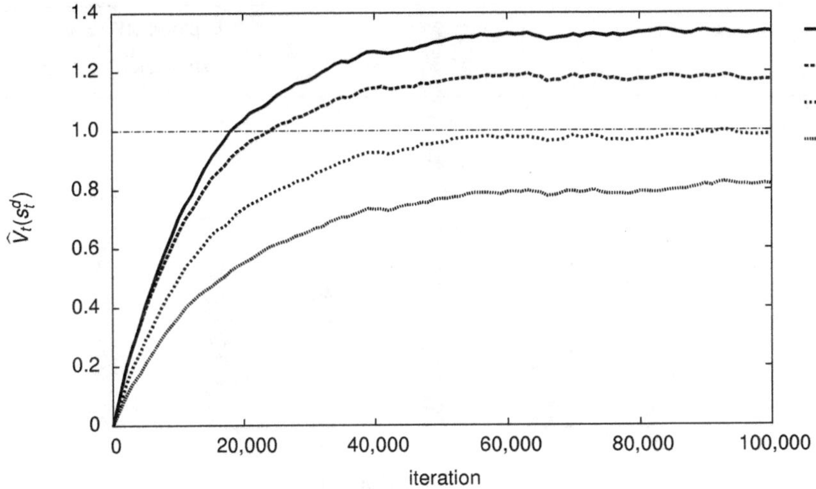

Fig. 7.4 Evolution of the value estimates for state $s_t^d = (0,0,3,0)$.

None of the four value estimates shown in Fig. 7.4 converges over the course of 200,000 iterations. Instead the estimates end up oscillating at an amplitude of approximately 0.01 each. The oscillation is not a problem with respect to s_1, s_2 as well as s_4, because the corresponding value estimates satisfy the conditions for an optimal policy in spite of their changing values. Unfortunately this is not true with respect to s_3. As the true value $V_3(s_3)$ is only slightly more than 1.0 and as $\hat{V}_3(s_3)$ seeks to achieve this value, the lack of precision due to oscillation causes a periodic change of the quality of the resulting policy.

Most likely, the major source of the oscillating behavior is the fact that Procedure 8 is executed subject to a constant stepsize. Thus, in order to achieve convergence, it suggests itself to run the procedure subject to declining stepsizes as induced for instance by the rules discussed in Sect. 5.2.2. Yet, for the problems treated in this section, the constant stepsize rule is considered to be sufficient as it leads to an optimal policy.

Part of the explanation of the differing performances of exploration and exploitation has already been discussed in the previous section. And indeed, the experimental results show that in the course of the first 100,000 iterations the exploration variant of Procedure 8 updates the values of a total of 405 different states, while its exploitation counterpart only reaches a total of 250 different states. Figure 7.5 offers more insights into the different behavior of the two approaches. It shows the relative frequencies of six different groups of states subject to either exploration or exploitation. States are grouped according to the number of updates they receive throughout the first 100,000 iterations. Only those states receiving at least one update are considered.

The figure shows that 18.8% of the states whose value estimates are updated by exploitation get only between one and ten updates. Of course the quality of a state

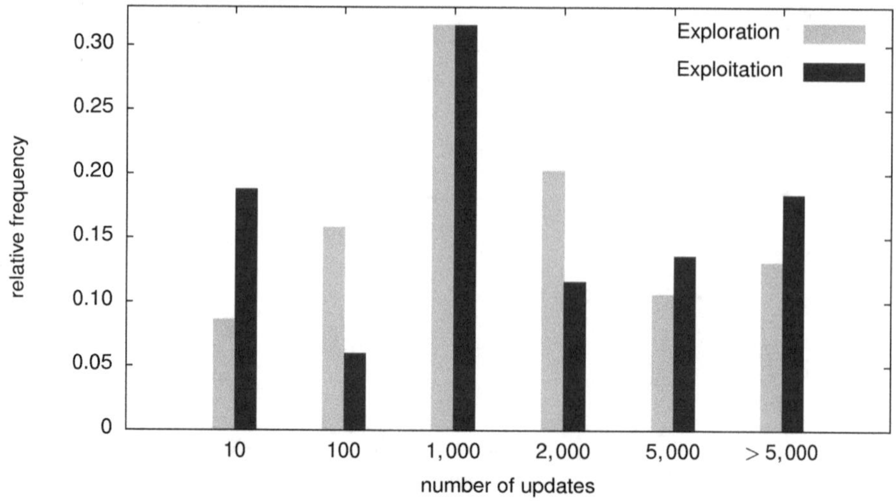

Fig. 7.5 Relative frequencies of states receiving a number of updates that is within one of the intervals $[1,10]$,$[11,100]$,$[101,1000]$,$[1001,2000]$,$[2001,5000]$ and $[5001,\infty)$.

value estimate generated from a given number of samples depends on the variance present in the data, which is induced by both the exogenous process and policy changes. Nevertheless, it seems quite likely that a maximum of ten updates often leads to an estimate of inferior quality. Moreover, it also seems quite likely that the value of some of the 18.4% of states which exploitation updates more than 5,000 times, could also have been estimated sufficiently well by use of a smaller number of samples. On the contrary, exploration has only 8.6% of the updated states within the category of one to ten updates. Thus, a smaller fraction of states runs the risk of having a relatively poor estimate. In addition, the fraction of states within the two categories of "more than 5,000 updates" and "between 2,001 and 5,000 updates" is relatively small if exploration is applied. That is, compared to exploitation, a lower fraction of states runs the risk of receiving more updates than necessary for obtaining a fairly good estimate. Instead, exploration shows a relatively high fraction of states receiving between 11 and 100 updates or receiving between 1,001 and 2,000 updates. Thus, in addition to the fact that, compared to exploitation, more states are updated by exploration, the latter is also characterized by distributing the total number of samples more uniformly among the states.

7.1.3 Limited Effectiveness

The previous section shows that perfect anticipation works well for dynamic routing of a service vehicle with three customer locations. In general, however, perfect anticipation suffers from limited effectiveness. As discussed in Sect. 3.4, major

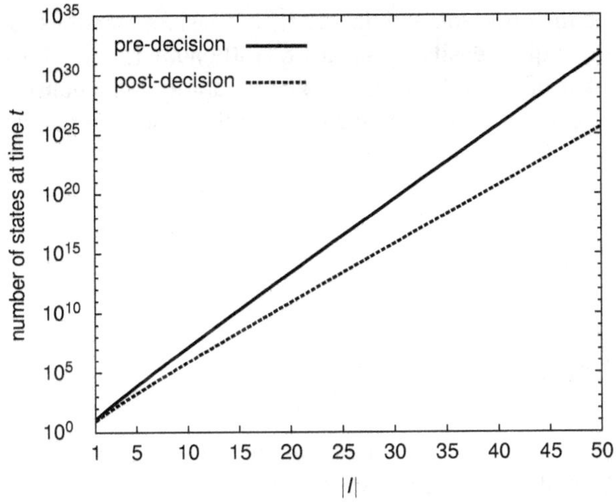

Fig. 7.6 The total number of states that may occur at an arbitrary point in time t.

difficulties arise due to the dimensionalties of both state space and decision space. Note that the additional challenge of evaluation of expected values has already been overcome by formulation of Bellman's equations around the post-decision state variables. More precisely, the post-decision formulation derived in Sect. 6.3.1 entails two very valuable consequences. The tremendous reduction of the computational burden that is enabled by the omission of expected values is accompanied by a significant reduction of the size of the state space. As a customer $i \in I$ may never be in state $r_{ti} = 1$ immediately after a decision has been made, the overall size of the state space reduces from $(\mathscr{T}+1) \times (|I|+2) \times 4^{|I|}$ down to $(\mathscr{T}+1) \times (|I|+2) \times 3^{|I|}$ if post-decision states are considered.

Figure 7.6 illustrates the state space dimensionality at an arbitrary point in time t for both the pre-decision state variables and the post-decision state variables. The more customer locations a problem instance comprises, the bigger becomes the gap between the numbers $|S_t|$ and $|S_t^d|$ of pre-decision states and post-decision states at time t. The small instance of Sect. 6.3.2 features $|I| = 3$, which results into $|S_t| = 320$ and $|S_t^d| = 135$. However, increasing the number of customers for example to $|I| = 50$ implies $|S_t| \approx 10^{32}$ in contrast to a number $|S_t^d|$ of "only" about 10^{25} post-decision states at one single point in time t.

In spite of the benefits gained from making use of post-decision states, a realization of perfect anticipation quickly becomes prohibitive as the number of customers increases beyond $|I| = 3$. Although some states will usually be relevant only in theory, the potential $(\mathscr{T}+1) \times |S_t^d|$ states to be handled are likely to imply an overwhelming computational burden already in the case of $|I| = 10$ for example. Additionally, the explosion of the computational burden may become much bigger, because a larger number of customers leads to an increase of the set of potential decisions. If $|I| = 3$, both the sets $\bar{R}_t$ of customers that can be confirmed at t without a violation of the time horizon and the set $\bar{M}_t$ of feasible locations to move to next, may be determined quite easily by means of full enumeration. However, as $|I|$ increases the number of possible decisions grows rapidly. As a consequence, merely verifying the feasibility of one single candidate decision may result in a computational task of nonpolynomial complexity.

In view of the limited effectiveness of perfect anticipation, the next section proposes approaches to approximate anticipation for dynamic routing of a service vehicle.

7.2 Approximate Anticipation

The following approach to approximate anticipation for dynamic routing of a service vehicle falls under the general actor-critic framework as proposed in Sect. 5.3. In particular, it relies on value function approximation by means of regression. Although such a high-level categorization of the approach is accomplished quite easily, the realization of approximate anticipation is by no means straightforward. A number of crucial questions must be answered before e.g. Procedure 9 may be set into operation. As shown in Sect. 5.3, an actor-critic method for approximate anticipation operates on the basis of both optimization and data mining. Thus, identification of the whole set of elements constituting the methodology of Sect. 4.1.3 is required.

Section 7.2.1 covers value function approximation, which — as shown in Sect. 5.1 — basically relies on adequate data mining operations. In contrast, Sect. 7.2.2 adresses the question of how to derive a decision model subject to a given system state. Section 7.2.3 then proposes different decision model approximation strategies and elaborates on their impact on the realization of approximate anticipation. The full scope of the resulting approach to approximate anticipation for dynamic routing of a service vehicle is illustrated in Sect. 7.2.4.

7.2.1 Value Function Approximation

As a first step towards value function approximation, the problem of dynamic routing of a service vehicle is considered from a system perspective. The elements of a refined system model (cf. Formula 4.2 in Sect. 4.1.1) are summarized in Fig. 7.7. This model comprises five types of objects. Four of these were already present in the basic system model illustrated in Fig. 6.1. In the refined system model, the basic object types "vehicle", "customer", "depot" and "road link" are joined by an object representing a decision making agent. Both the variable and the constant attributes of customers, depots as well as road links are identical to those appearing in the basic system model. Merely the vehicle is represented at a slightly more detailed level. The basic model of Fig. 6.1 does not take into account the fact that in the strict sense the amount $\mathscr{T}$ of time available is an attribute of the vehicle. At first glance,

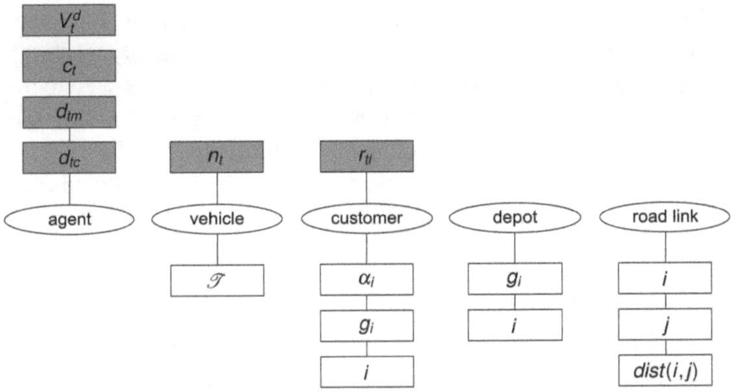

Fig. 7.7 Refined system model of dynamic routing of a service vehicle. System objects are denoted as *ellipses* while their attributes appear as *boxes*. Variable attributes are represented as *gray-shaded boxes* and constant attributes appear as *white boxes*.

the time available may seem to be an attribute of each of the system objects. Yet, on closer inspection it turns out to belong to the vehicle only, as, from an economic perspective, the vehicle is a resource to be utilized and consequently $\mathscr{T}$ determines the overall resource capacity.

The decision making agent is defined by four variable attributes. Two of these represent the decisions made while another two reflect an evaluation of these decisions. As defined in Sect. 6.3.1, the agent's decision at time t resolves into a confirmation operation d_{tc} and a vehicle move operation d_{tm}. Moreover, the evaluation attributes are given by the contribution c_t as well as the value V_t^d of the post-decision state s_t^d at time t. Note that the elements of a post-decision state $s_t^d = (n_t^d, r_{t1}^d, r_{t2}^d, \ldots, r_{t|I|}^d)$ must not be represented explicitly in the system model, as they result implicitly from the pre-decision state $s_t = (n_t, r_{t1}, r_{t2}, \ldots, r_{t|I|})$ and the decision $d_t = (d_{tc}, d_{tm})$ at time t.

In summary, at time t an appearance σ_t of the system that is established by dynamic routing of a service vehicle is represented by a total of 15 attributes. Besides two evaluation attributes and two decision attributes, σ_t comprises a number of variable system attributes given by the current position of the vehicle and the customers' request states. The remaining regular attributes of σ_t are constant. At a point in time t the value of any attribute except V_t^d can be observed. Thus, in terms of Sect. 4.1, information about system structure is required in order to derive adequate surrogates for V_t^d. Part of the system structure is represented by the value functions $\forall t \in \mathscr{T} : V_t^d : S_t^d \mapsto \mathbb{R}$. Information about these static system relations may be gained by the approaches proposed in Sect. 5.1.2. Yet, value function approximation requires hypotheses about system structure. Such hypotheses may be formulated based on the following concepts and quantities, each of which can be extracted from a system appearance σ_t at time t:

- *planned route*: The planned route x_t^d is an ordered set whose elements are given by the customers $\{i | r_{ti}^d = 2 \vee i = n_t^d\}$ as well as the end depot E. It imposes a sequence on these elements and represents the current plan for serving every confirmed customer while traveling from n_t^d to the end depot, i.e., the first element of x_t^d is equal to n_t^d, while the last element is equal to E.
- *route length*: The quantity $l(x_t^d)$ represents the length of the planned route x_t^d. The length is equal to the total of those distances $dist(\cdot,\cdot)$, that are associated with the road links connecting the geographic locations of the elements of x_t^d according to the order imposed by x_t^d.
- *presumed deviation*: The quantity $pd_t(i, x_t^d)$ is the product of the request probability α_i of customer i and the minimum additional distance to be traveled, if i must be inserted into x_t^d. The presumed deviation is of relevance only with respect to customers that did not request for service yet. Thus, if at minimal cost such a customer i is inserted between two subsequent elements $j, k \in x_t^d$, the presumed deviation turns out to be
$$pd_t\left(i, x_t^d\right) = \alpha_i\left(dist(j,i) + dist(i,k) - dist(j,k)\right) .$$
- *slack*: The quantity $sl_t(x_t^d)$ is the surplus of distance that may be covered by the vehicle subject to the planned route x_t^d. Assuming a travel speed of one unit of time per one unit distance as well as letting $dist(n_t, n_t) = 1$ leads to
$$sl_t(x_t^d) = \mathcal{T} - t - dist(n_t, d_{tm}) - l(x_t^d) .$$
- *presumed number of requests*: The quantity pr_t represents the number of requests, that would be expected to occur throughout the time horizon, if only those customers were considered that did not issue a request until the current point in time t. Thus, the presumed number of requests can be derived from the requests probabilities α_i as
$$pr_t = \sum_{i | r_{ti} = 0} \alpha_i .$$

Given a policy π, the true value $V_t^{d,\pi}(s_t^d)$ of a post-decision state s_t^d at time t may be expressed in terms of accumulated contributions $C_{t'}(s_{t'})$ of successor states $s_{t'}$, i.e.,

$$V_t^{d,\pi}(s_t^d) = \mathbb{E}\left[C_{t'}(s_{t'})\right] .$$

In general, an accumulated contribution $C_{t'}(s_{t'})$ is – to a certain extent – determined by the number of customers that issues a request within $(t \ldots \mathcal{T}]$. Consequently, there is an influence of the number of these customers on the value $V_t^{d,\pi}(s_t^d)$. At time t, an upper bound of the number of future requests is given by the number of customers with $r_{ti} = 0$. Additionally taking into account the customers' request probabilities suggests that there is a correlation between the presumed number of requests pr_t and the post-decision state value $V_t^{d,\pi}(s_t^d)$. In the special case of $pr_t = 0$, the presumed number of requests fully determines $V_t^{d,\pi}(s_t^d)$, i.e., $(pr_t = 0) \Rightarrow (V_t^{d,\pi}(s_t^d) = 0)$. In many other cases, a larger value of pr_t is likely to result into more future confirmations, which implies a larger value of the post-decision state $s_{t'}$.

However, the correlation between pr_t and $V_t^{d,\pi}(s_t^d)$ may be weak. A larger value of pr_t will not result into an increase of $C_{t'}(s_{t'})$ without the capability of confirming additional requests. At time t this capability is determined by both the available surplus of travel time and the travel time that is required for visiting additional customer locations. While the former is represented by the slack sl_t at t, an approximation of the latter may be gained by summing up the presumed deviations pd_t of those customers that did not request for service until t. Both a larger amount of slack and a smaller total of presumed deviations are likely to increase the capability of making additional confirmations.

Bringing together the influence of the presumed number of requests and the influence of the capability of making additional confirmations gives rise to the following hypothesis.

Hypothesis 1: The higher the ratio of slack and average presumed deviation turns out to be at time t, the larger is the expected accumulated contribution at the following decision time. Any one unit increase of an arbitrary value of this ratio at time t implies an increase of the expected accumulated contribution by the same fixed amount. Thus, at time t

$$\mathbb{E}\left[C_{t'}(s_{t'})\right] \propto \frac{sl_t}{\frac{\sum_{i|r_{ti}=0} pd_t}{pr_t}} = \frac{sl_t}{\sum_{i|r_{ti}=0} pd_t}\, pr_t \,. \tag{7.3}$$

Note that the right hand side of Eq. 7.3 yields customer as unit. Thus, the ratio $\frac{sl_t}{\sum_{i|r_{ti}=0} pd_t}\, pr_t$ may be considered as the *presumed number of confirmed customers*, in opposition to the expected number of confirmed customers denoted by the leftmost expression of Eq. 7.3.

Hypothesis 1 may be consulted for a value function approximation that is based on $g_t(s_t^d) = (g_\sigma^0(\sigma_t))$, with[4]

$$g_\sigma^0(\sigma_t) = \frac{sl_t}{\sum_{i|r_{ti}=0} pd_t}\, pr_t \,. \tag{7.4}$$

Letting $r_t = (r_t^0)$ and imposing an information structure

$$\widetilde{V}_t^d(r_t, g_t(s_t^d)) = r_t^0\, g_\sigma^0(\sigma_t) \tag{7.5}$$

leads to a total of $\mathscr{T}$ different approximate value functions. This kind of value function approximation makes use of a single preprocessing operation $g_\sigma^0(\sigma_t)$.[5] A value returned by $g_\sigma^0(\sigma_t)$ relies on the slack, the presumed number of requests as well as the presumed deviations extracted from the system appearance σ_t at time t. As postulated by Hypothesis 1, a value function in the sense of Eqs. 7.4 and 7.5 assumes

---

[4] Note that s_t^d is fully determined by s_t and d_t, which are both integral components of σ_t.

[5] According to the types of preprocessing operations introduced in Sect. 4.1.2, $g_\sigma^0(\sigma_t)$ falls into the category of attribute construction.

the same marginal utility r_t^0 of a presumed confirmation for any state $s_t^d \in S_t^d$ that may occur at time t.

While this assumption applies to many situations, there also are situations in which it proves to fail. Consider the *Move early*-scenario of Sect. 6.3.2 as an example. In this scenario, the requesting customer 1 receives a confirmation and the vehicle could either remain at g_S or move on to g_1 immediately. The discussion in Sect. 6.3.2 shows that moving on to g_1 increases the likelihood of an additional confirmation. Yet, applying $g_6(\cdot)$ to both the post-decision state $s_6^{d,m} = (1,3,0,0)$ resulting from decision $d_6^m = (\{1\},1)$ and the post-decision state $s_6^{d,w} = (0,2,0,0)$ resulting from decision $d_6^w = (\{1\},0)$ produces exactly the same numbers of presumed confirmations. At this point, an approximate value function $\widetilde{V}_6^d$ derived from Eq. 7.5 ignores the fact that the utility of a presumed confirmation is higher if the vehicle moves on to g_1 instead of waiting at g_S. Consequently, the following refinement of Hypothesis 1 may be considered.

Hypothesis 2: The higher the ratio of slack and average presumed deviation turns out to be at time t, the larger is the expected accumulated contribution at the following decision time. Subject to the current vehicle position, *a one unit increase of an arbitrary value of this ratio at time t implies an increase of the expected accumulated contribution by the same fixed amount. Thus,* for each possible vehicle position n_t^d *at time t,*

$$\mathbb{E}\left[C_{t'}(s_{t'})\right] \propto \frac{sl_t}{\frac{\sum_{i|r_{ti}=0} pd_t}{pr_t}} = \frac{sl_t}{\sum_{i|r_{ti}=0} pd_t}\, pr_t \;.$$

Hypothesis 2 represents an alternative to Hypothesis 1 and may be consulted for a value function approximation that is based on $g_t(s_t^d) = \left(g_\sigma^0(\sigma_t),\, g_\sigma^1(\sigma_t)\right)$, with

$$g_\sigma^1(\sigma_t) = n_t^d \;. \tag{7.6}$$

An information structure allowing for different marginal utilities $r_t = (r_t^0, \ldots, r_t^{|I|})$ of a presumed customer with respect to different post-decision vehicle positions n_t^d at time t, is then given by

$$\widetilde{V}_t^d\left(r_t, g_t(s_t^d)\right) = r_t^n\, g_\sigma^0(\sigma_t) \quad \text{if } g_\sigma^1(\sigma_t) = n \;. \tag{7.7}$$

This structure generates $\mathscr{T}$ different approximate value functions, with a single value function comprising $|I|+1$ parameters r_t^n. Naturally, the more fine-grained Hypothesis 2 is reflected in a larger number of parameters to be determined. As a consequence, the computational effort for deriving the approximate value functions from Eq. 7.7 is likely to be significantly higher than the effort implied by Eq. 7.5.

7.2.2 Decision Model Identification

Selection of a type of value function approximation determines the structure of the objective functions for decision making within an actor-critic method. In the case of dynamic routing of a service vehicle, the structure of the full decision model at a time t comprises one of the value function approximations introduced in the previous section and the generic set of feasible decisions given by Eq. 6.1 in Sect. 6.3.1. As a result, the optimization problem to be solved at t is of type

$$P_t^{\prime} = \left(\left\{ d_t | d_{tc} \in \bar{R}_t \wedge d_{tm} \in \bar{M}_t \right\},\ |d_{tc}| + \widetilde{V}_t^d (r_t, g_t(s_t^d)) \right) , \tag{7.8}$$

with $\bar{R}_t$ being the sets of requesting customers that may be confirmed without a violation of the time horizon and $\bar{M}_t$ being the permitted vehicle moves subject to the confirmation decision d_{tc}.

Two steps are required for deriving a decision model from Eq. 7.8. On the one hand the parameters r_t of the approximate value function must be set. On the other hand, the candidate decisions making up the set of feasible decisions at t must be identified. A candidate decision $d_t = (d_{tc}, d_{tm})$ is feasible, if a planned route x_t^d with $sl_t(x_t^d) \geq 0$ exists. Thus, feasibility can be taken for granted if a route satisfying this condition can be specified. One approach to checking the feasibility of a candidate d_t consists in solving the optimization problem

$$P_t^x = (X_t^d, sl_t(x_t^d)) , \tag{7.9}$$

with X_t^d being the set of planned routes that may be constructed at time t subject to d_t. Solving Problem P_t^x by means of an exact algorithm returns a planned route $x_t^{d,\star}$ and according to Eq. 2.1,

$$\forall x_t^d \in X_t^d : sl_t(x_t^{d,\star}) \geq sl_t(x_t^d) . \tag{7.10}$$

In general, the availability of $x_t^{d,\star}$ is both necessary and sufficient for making a definite statement about the feasibility of d_t. Clearly, feasibility is granted if $sl_t(x_t^{d,\star}) \geq 0$, while d_t is for sure infeasible in case $sl_t(x_t^{d,\star}) < 0$. However, finding $x_t^{d,\star}$ may be prohibitive, as the optimization problem of Eq. 7.9 represents a variant of the traveling salesman problem (cf. Sect. 6.1.1). The latter belongs to the class of NP-complete problems, which means that an exact algorithm that guarantees a global optimum within polynomial runtime cannot be provided up to now.

Fortunately, in many cases a definite statement about the feasibility of d_t can be made by considering a route $x_t^d \in X_t^d$ with $x_t^d \neq x_t^{d,\star}$. Insofar as x_t^d implies $sl_t(x_t^d) \geq 0$, the candidate decision d_t can for sure be categorized as feasible. Thus, a non-optimal solution to Problem 7.9, which can be specified at a relatively low computational cost, may be sufficient. As discussed in Sect. 2.2.2, such a solution can be generated by means of either heuristics or metaheuristics. Like most optimization techniques, the GRASP[6] metaheuristic (Feo and Resende, 1989, 1995) has been successfully

[6] GRASP stands for "greedy randomized adaptive search procedure".

applied in many different contexts (Festa and Resende, 2001). Moreover, GRASP shows nice scaling properties, allowing for the method to either degenerate into a normal heuristic or to gradually unfold the qualities of a metaheuristic at the expense of a higher computational effort. For these reasons, GRASP is used for checking the feasibility of a candidate decision in the following.[7] The resulting Procedure 11 requires a candidate decision d_{τ_k} and a planned route $x^d_{\tau_{k-1}}$ as input.

Procedure 11: GRASP(d_{τ_k}, $x^d_{\tau_{k-1}}$)

```
ε ← 0
for number of GRASP-iterations do
    x ← setBaseRoute(x^d_{τ_{k-1}}, d_{τ_k m})
    x ← ConstructGreedyRandomizedSolution(x, d_{τ_k c})
    x ← LocalSearch(x)
    if sl_{τ_k}(x) ≥ ε then
        x^d_{τ_k} ← x
        ε ← sl_{τ_k}(x^d_{τ_k})
    end
end
```

The procedure consists of four major steps within each iteration. First, a basic route x is derived from the planned route $x^d_{\tau_{k-1}}$ and the move operation $d_{\tau_k m}$. The basic route contains each customer that was in state $r^d_{\tau_{k-1} i} = 2$. Additionally the first element of the route is set equal to $d_{\tau_k m}$ while the last element is given by the end depot E.

In the second step, the remaining elements of $d_{\tau_k c}$ are inserted into the basic route by application of a randomized construction heuristic. This heuristic builds a restricted candidate list of the most suitable elements of $d_{\tau_k c}$ to be inserted into x. Suitability is determined by an insert-savings[8] criterion, i.e., for each customer $i \in d_{\tau_k c}$ the difference between the length of the route $(0, i, E)$ and the minimal deviation occuring from insertion of i into x is determined. The customer showing the minimal difference is considered most suitable and ranks first in the candidate list. One of the candidates from this list is then selected randomly for actual insertion. Subsequently a new restricted candidate list is established for insertion of the next element into the resulting route. The construction heuristic terminates as soon as each of the elements of $d_{\tau_k c}$ is included into the route.

In the next step, x is improved by a local search technique. This technique operates on a 1-shift neighborhood (Bertsimas and Howell, 1993) for iteratively generating a route that represents a local optimum as defined in Eq. 2.2. For computational

[7] Nevertheless, GRASP is of course not the only optimization technique that can be applied at this point. Suitable alternatives are for example genetic algorithms and simulated annealing.

[8] The insert-savings criterion is derived from the traditional savings method for route construction (Clarke and Wright, 1964). A detailed description of insert-savings and other routing heuristics is provided by, e.g., Kilby et al (2000).

reasons, it is implemented as a first-improvement local search, i.e., the next iteration is started as soon as a better solution is found within the neighborhood.

At the end of each GRASP-iteration the route x is considered as a candidate for the current plan $x^d_{\tau_k}$, if $sl_{\tau_k}(x) \geq 0$ and if at the same time $sl_{\tau_k}(x)$ is greater or equal to the slack of the shortest route found so far. Strictly speaking, feasibility of d_{τ_k} can be taken for granted as soon as $sl_{\tau_k}(x) \geq 0$ occurs for the first time. Nevertheless, it may be beneficial to get on with Procedure 11 after d_{τ_k} is already known to be feasible, as this may lead to a planned route which provides more slack than the first route satisfying $sl_{\tau_k}(x) \geq 0$. Moreover, an upper limit of the number of GRASP-iterations must be selected, in order to ensure termination of Procedure 11. Thus, it might be that a feasible decision d_{τ_k} is considered infeasible because no route x with $sl_{\tau_k}(x) \geq 0$ is found within the given number of iterations.

Note that a special case occurs at decision time $\tau_0 = 0$. At τ_0 the number of late requesting customers actually requesting for service equals zero. The vehicle may either move to one of the early requesting customers or wait at g_S. Subject to $d_{\tau_0 c} = \emptyset$ and $d_{\tau_0 m} \in I_E \cup 0$, an initial planned route $x^d_{\tau_0}$ may be determined by running Procedure 11 with $d_{\tau_0 c} = I_E$, $d_{\tau_0 m} \in I_E \cup 0$ and $x^d_{-1} = (0, E)$.

7.2.3 Decision Model Approximation

If a sufficient number of GRASP-iterations is executed per candidate decision, the method of the previous section leads to a decision model satisfying Eq. 7.8. Yet, in practice the need for computational efficiency suggests permitting a limited number of iterations only. As a consequence, the method is likely to generate an approximation of the set of feasible decisions. The computational burden that results from running Procedure 11 for each candidate decision d_t can be controlled to a certain extent via the number of GRASP-iterations. However, even if a small number is selected, Procedure 11 still must be executed $|M_t|$ times for each $d_{tc} \in \mathscr{P}(R_t)$. A further reduction of the computational effort required for establishing a decision model at time t can be realized by a stronger type of approximation of the set of feasible decisions. A suitable approximation can be derived from the following hypothesis.

Hypothesis 3: Let d_{tc} be a candidate set of confirmations at time t. Then, in general the most important tradeoff subject to a confirmation of d_{tc} is whether the vehicle should wait at its current location n_t or move on to another geographical location.

The reasoning behind Hypothesis 3 starts from the fact that candidate decision $d^w_t = (d_{tc}, n_t)$ always implies a planned route comprising one more road link than any other candidate decision subject to d_{tc}. As a higher number of road links to be traveled in the future is likely to decrease the sum of the presumed deviations pd_t, the wait operation shows advantages compared to alternative move operations. Moreover, waiting may be conducted subject to alternative choices of the successor location. Thus, the planned route x^d_t generated by Procedure 11 for checking the feasibility of d_t^w potentially exceeds the slack implied by alternative move operations, as the latter enforce a specific successor location each. Taking into account Eq. 7.3, both the potentially low sum of presumed deviations and the potentially large amount of slack suggest d_t^w to lead to a successor state of high value. Candidate decision $d_t^w = (d_{tc}, n_t)$ is considered feasible, if Procedure 11 generates a planned route x_t^d with $sl_t(x_t^d) \geq 0$. At the same time, this proves the feasibility of candidate decision $d_t^m = \left(d_{tc}, x_t^d(2)\right)$, with $x_t^d(2)$ being the second element of x_t^d. Note that subject to a sufficient number of iterations of Procedure 11, $x_t^d(2)$ implies the maximal slack among all the move operations $M_t \backslash \{n_t\}$. Thus, a successor state achieved by d_t^m is likely to be more valuable than the sucessor states resulting from a move to an alternative element of $M_t \backslash \{n_t\}$. Of course, Eq. 7.3 and the associated discussion about state values show that this assumption may fail.[9] Nevertheless contrasting d_t^w with d_t^m is promising as in general they both show advantages over the remaining move operations subject to d_{tc}. In addition, both the *Wait*-Scenario and the *Move early*-Scenario of Sect. 6.3.2 may be considered as an endorsement of the great importance of the tradeoff between waiting and moving on.

Against the background of the preceding discussion, Hypothesis 3 may be consulted for approximating the set of feasible decisions $D_t(s_t)$. Both d_t^w and d_t^m are promising candidate decisions that focus on different characteristics of the successor state. Considering only those two movement operations per candidate set d_{tc} leads to an approximate set $\widetilde{D}_t(s_t)$ of feasible decisions. This set can be specified as $\widetilde{D}_t(s_t) = \cup_{d_{tc} \in \mathscr{P}(R_t)} \widetilde{D}(d_{tc}, x_t^d)$ with $a = dist\left(n_t, x_t^d(2)\right)$ and

$$\tilde{D}(d_{tc}, x_t^d) = \begin{cases} \left\{d_t | d_{tm} \in \left\{n_t, x_t^d(2)\right\}\right\} & : \quad sl_t(x_t^d) \geq 0 \\ \left\{d_t | d_{tm} = x_t^d(2)\right\} & : \quad -a \leq sl_t(x_t^d) \leq -1 \\ \emptyset & : \quad sl_t(x_t^d) < -a \end{cases} . \tag{7.11}$$

As a result, an optimization problem resulting from Eq. 7.8 may be replaced by a problem of type

$$P_t'' = \left(\widetilde{D}_t(s_t),\ |d_{tc}| + \widetilde{V}_t^d\left(r_t, g_t(s_t^d)\right)\right) . \tag{7.12}$$

Substituting Problem 7.12 for Problem 7.8 raises the level of approximation. For each d_{tc}, only two elements out of the set of post-decision states are selected by means of the planned route x_t^d. As these states are assumed to show a high value, the set of feasible decisions is reduced accordingly. Thus, Hypothesis 3 may be considered as a quite coarse approximation of the post-decision value function, to which the decision maker reacts by adapting the set of feasible decisions. In terms of Sect. 2.3.2 this process introduces an implicit type of anticipation. Consequently,

[9] Depending on the geographical locations of the customers that did not request for service yet, conducting the move operation with maximal slack may lead to an increase of the sum of presumed deviations.

the pure approximate anticipation, as realized by making decisions according to Eq. 7.8, is reduced in quality.

Making decisions according to Eq. 7.12 may be considerd as a hybrid of approximate and implicit anticipation. This hybrid, however, entails a significant reduction of computational cost. While pure approximate anticipation enforces executing Procedure 11 a total of $| M_t |$ times per d_{tc}, the combination of approximate and implicit anticipation only requires a single run per d_{tc}.

7.2.4 The Full Scope of the Approach

Sections 7.2.1 through 7.2.3 provide the means for realizing approximate anticipation for dynamic routing of a service vehicle. The elements introduced up to this point define a whole family of actor-critic methods that may be subsumed under the framework defined in Sect. 5.3 and outlined in Fig. 5.1. The full scope of this approach is summarized in the following.

- *Information structure.* Two alternative information structures are provided for value function approximation. The structure of Eq. 7.5 is derived from Hypothesis 1 and represents a linear approximation of the true value function. A more fine-grained piecewise linear approximation is established by the structure of Eq. 7.7, which is derived from Hypothesis 2.
- *Value estimates.* Each actor-critic method requires selection of initial state value estimates. These are determined by the initial values of the parameters of the value functions. In the absence of converse information, every parameter may be initialized with 0. However, different initial settings will lead to a different behavior of an actor-critic method. In addition, different actor-critic methods result from different types of sample value estimates to be considered for carrying out value function updates. Each of the TD(λ) approaches introduced in Sect. 3.2.3 may be considered.
- *Stepsize rule.* The behavior of an actor-critic method is also heavily characterized by the selected stepsize rule. A discussion of the broad range of relevant stepsize rules is provided in Sect. 5.2.2.
- *State sampling.* The influence of state sampling on the behavior of an actor-critic method is illustrated in Sect. 7.1. Both exploitation sampling and exploration sampling are proposed in this section. However, intermediate types of state sampling may be considered as additional alternatives.
- *Decision model structure.* Two alternative decision model structures are proposed. On the one hand, a higher level of anticipation may be realized by establishing decision models as proposed by Problems 7.8. On the other hand, a significant reduction of the computational burden may be realized by deriving decision models from Problems 7.12 at the expense of a lower degree of anticipation.
- *Decision model identification.* Verification of the feasibility of a candidate decision requires solving Problem 7.9 by means of Procedure 11. Moreover, the

resulting planned route is part of the criterion of Problems 7.8 and Problems 7.12 respectively. Thus, the behavior of an actor-critic method depends on the characteristics of Procedure 11. The latter may be configured by the number of GRASP-iterations as well as by the number of elements of the restricted candidate list. Moreover the procedure may be conducted either with or without local search, which may itself be configured by a variety of different neighborhoods.

Each of the preceding characteristics represents an important design decision with respect to approximate anticipation for dynamic routing of a service vehicle. Making these decisions implies a tradeoff between computational burden and level of anticipation. Nevertheless, any of the resulting methods takes into account the nature of a dynamic decision problem by falling into the category of generalized actor-critic methods as defined in Sect. 5.3.

However, adhering to the actor-critic framework does not necessarily lead to more effective decisions. Possibly, the positive effect of the actor-critic principle is blurred by the hypothetical character of the downgrade from perfect anticipation to approximate anticipation. An assessment of the latter may be achieved by additionally considering approaches realizing a lower degree of anticipation. For the case of dynamic routing of a service vehicle such approaches are presented in the next section.

7.3 Non-reactive Anticipation

The following sections propose non-reactive approaches to anticipatory optimization for dynamic routing of a service vehicle. State values are not considered. Instead, decisions are derived from general hypotheses about the remaining decision process. The approaches of Sect. 7.3.1 take into account the customers' request probabilities α_i for hypothesis generation. In contrast, the approaches of Sect. 7.3.2 rely on hypotheses that are formulated without probabilistic information. Every approach presumes an initial planned route, that is determined by running Procedure 11 subject to the arguments $d = (I_E, 0)$ and $x = (0, E)$.

7.3.1 Probabilistic Approaches

Three probabilistic approaches to non-reactive anticipation are provided in the following. The *Presumed Confirmations* (PC) approach relies on the presumed number of confirmed customers as introduced in Sect. 7.2.1. Both the *Longest Wait* (LW) approach and the *Closest Wait* (CW) approach are related to approaches from the literature (Thomas, 2007) and rely on the concept of center of gravity (e.g., Nahmias, 2008).

- *Presumed Confirmations.* PC uses both the contribution $|d_{tc}|$ of a decision d_t and the presumed number of confirmed customers entailed by d_t as evaluation criteria. At any decision time the feasible decision maximizing the sum of actually confirmed customers and presumed confirmed customers is executed. Just as with the approach of Sect. 7.2, PC may either consider the complete set of possible move operations or restrict the move operations as proposed in Sect. 7.2.3. In any case the behavior of PC depends on the configuration of Procedure 11.
Note that PC may be interpreted as considering the value of a successor state to be equal to the presumed number of confirmed customers, i.e., with respect to Eq. 7.5, PC imposes approximate value functions $\widetilde{V}_t^d(g_t(s_t^d)) = 1 * g_\sigma^0(\sigma_t)$.
- *Longest Wait.* At any decision time, LW confirms as many new requesting customers as possible. Confirmation of a set $d_{tc} \in \mathscr{P}(R_t)$ is possible, if a planned route x_t^d comprising d_{tc} can be specified such that $sl_t(x_t^d) \geq -a$ (cf. Eq. 7.11). Thus, a maximum number of $|\mathscr{P}(R_t)|$ runs of Procedure 11 may be required until the largest possible subset d_{tc} is identified.[10]
If the confirmation operation implies $-a \leq sl_t(x_t^d) \leq -1$, the vehicle moves on to $x_t^d(2)$. Otherwise the vehicle waits instead of moving to $x_t^d(2)$ only if n_t is the location within x_t^d, that allows for the latest allowable departure time. The departure time associated with an element i of x_t^d is defined as the latest time at which the vehicle may leave g_i for going to a specific location g_{cog} before moving from g_{cog} to g_E via i's successors as given by x_t^d.
The location g_{cog} represents the "center of gravity" of the set $\bar{I}_L$ of those late requesting customers that did not request for service up to time t and that at the same time may individually be inserted into x_t^d without causing a violation of the time horizon. It is defined as the point $g_{\mathrm{cog}} = (g_{\mathrm{cog}}^x, g_{\mathrm{cog}}^y)$, that minimizes the weighted square of the Euclidean distance between these late requesting customers' locations $g_i = (g_i^x, g_i^y)$. As proposed by Thomas (2007), the weight $\bar{\alpha}_i$ associated with customer i may be defined as i's request probability per time unit, which can easily be derived from α_i. Thus, the center of gravity is given by
$$g_{\mathrm{cog}}^x = \frac{\sum_{i \in \bar{I}_L} \bar{\alpha}_i g_i^x}{\sum_{i \in \bar{I}_L} \bar{\alpha}_i} \quad \text{and} \quad g_{\mathrm{cog}}^y = \frac{\sum_{i \in \bar{I}_L} \bar{\alpha}_i g_i^y}{\sum_{i \in \bar{I}_L} \bar{\alpha}_i}.$$
On the basis of x_t^d and g_{cog}, a customer i's departure time is allowable if it is later than or equal to the vehicle's earliest arrival time at i according to x_t^d.
- *Closest Wait.* CW is similar to LW with respect to d_{tc}, i.e., as many new late requests as possible are confirmed at each decision time. Moreover, as in the case of LW, the decision maker either decides to move on to $x_t^d(2)$ or to wait for one unit of time at n_t. However, in contrast to LW, CW waits at n_t only if n_t is closer to the center of gravity than any other element of x_t^d.

[10] Note that such a subset can only be identified subject to the configuration of Procedure 11, i.e., due to its heuristic nature the procedure may result in a subset whose cardinality is smaller than the cardinality of the truly largest possible subset of confirmations.

7.3.2 Conventional Approaches

The non-reactive approaches presented in the remainder of this section each work without probabilistic information. The *Randomized Wait* (RW) approach may decide to wait if the maximum number of confirmations allows for waiting. Both the Distributed Wait (DW) approach and the Wait at start (WS) approach are inspired from the literature (Thomas, 2007). DW allocates the same amount of waiting time at each confirmed customer and WS allocates the whole amount of waiting time at the start depot. Each of the three approaches makes the confirmation decision d_{tc} in the same way as LW and each of the approaches considers n_t and $x_t^d(2)$ as candidate decisions with respect to d_{tm}. The criteria for making the decision on whether to wait or not are as follows.

- *Randomized Wait.* If the slack of the planned route resulting from the confirmation decision d_{tc} allows for waiting, RW randomly decides on whether to wait or not. At each decision time, both n_t and $x_t^d(2)$ are assigned a 0.5 probability of being selected as actual move operation d_{tm}. RW gives in to the fact that in the absence of a criterion for making the decision of whether to wait or not, both options must be treated as if they were of equal quality.
- *Distributed Wait.* The total amount of waiting time available after making decision d_{tc} is equal to the slack of the planned route corresponding to d_{tc}. DW distributes this amount of waiting time equally among the elements of $x_t^d \setminus \{E\}$. The vehicle moves on to $x_t^d(2)$ either in case the waiting time assigned to n_t is less than one unit of time or in case the total amount of waiting time assigned to n_t has elapsed without any new late requests. The waiting time per routed customer is recomputed each time new late requests have been accepted.
- *Wait at Start.* The start depot is the only waiting position permitted by WS. The vehicle waits at g_S confirming every new late request until the slack of the planned route is equal to zero. It then moves from the start depot to the end depot according to x_t^d without waiting anywhere. Further customer requests are rejected.

Chapter 8
Computational Study

This chapter provides insights into the behavior of the different approaches to anticipatory optimization introduced in Sects. 7.2 and 7.3. The insights are derived from computational experiments with a set of diverse instances of the problem of dynamic routing of a service vehicle. Each of these instances is exposed to each of the approaches such that the quality of an approach can be understood with respect to both varying problem characteristics and the qualities of its competitors.

A particular focus of the study is on contrasting the impact of different degrees of anticipation. Starting from conventional non-reactive anticipation the experiments are extended to probabilistic non-reactive anticipation and to approximate anticipation. The latter is consulted in terms of both the elementary value function approximation and the fine-grained value function approximation introduced in Sect. 7.2.1. Both of these approximations are not only examined with respect to the resulting quality of solutions but also with respect to the solutions' properties.

A detailed description of the experimental setup considered is provided in Sect. 8.1. Subsequently, Sect. 8.2 presents results generated by non-reactive anticipation. In Sect. 8.3 these results are contrasted with the behavior of approximate anticipation in terms of the elementary value function approximation. Finally, Sect. 8.4 illustrates the consequences of substituting this approximation with its more fine-grained counterpart.

8.1 Experimental Setup

The number of possible instances of the problem of dynamic routing of a service vehicle is large. Moreover, such an instance may be treated by a vast number of configurations of the framework introduced in Sect. 7.2. For this reason, the following subsections identify a subset of instances and configurations to be consulted for computational experiments. Section 8.1.1 defines a number of problem instances capturing prominent characteristics of real-world scenarios. Then, Sect. 8.1.2

imposes presettings on the configuration of the actor-critic framework in support of both computational efficiency and comparability to non-reactive anticipation.

8.1.1 Problem Instances

The problem instances involved in the computational experiments consider the routing of one service vehicle over a horizon of $\mathcal{T} = 350$ time units. This corresponds to a driver's shift of approximately six hours, if one unit of time equals one minute. Throughout the driver's shift, the vehicle is assumed to operate in a service region comprising 49 geographical customer locations. Each customer location g_i is assumed to be reachable, i.e., it is assumed to be close enough to both the start depot and the end depot, such that $dist(0,i) + dist(i,E) \leq \mathcal{T}$.

The fraction of early requesting customers is one of the most defining properties of a problem instance. In a practical context, the number of early requesting customers may be the same at the beginning of every shift. As an example, consider the case of a fixed number of registered customers that always must be served. Similarly a transportation company may have to visit a fixed number of collection points, which are treated as the early requests of a shift. However, the fraction of early requesting customers may also be subject to fluctuation. Fluctuation typically occurs if the rejected late requests of today are considered as the early requests of tomorrow or if customers call in outside the time of the current shift.

In any case, the fraction of early requesting customers is an important problem characteristic. Thus, a variety of fractions are consulted for the computational experiments of the following sections. Either five, ten or twenty out of the 49 potential customers are assumed to issue early requests. These three settings result into early request ratios of approximately 19%, 34% and 58%, if half of the respective late requesting customers actually issue a request throughout the time horizon.

In a real-world scenario, the behavior of a late requesting customer is typically determined by a large variety of influences. In order to strike the balance between the many types of behaviors that may occur, each late requesting customer's request probability α_i is drawn from a normal distribution $\mathcal{N}(0.5, 0.01)$. As a consequence, approximately fifty percent of these customers will request for service throughout the period of 350 time units.[1]

Apart from the probabilities of late requests and the ratio of early requesting customers, the major characteristic of a problem instance is given by the geographical distribution of the customer locations within the service region. Different types of distribution may occur in practice. For example, a service vehicle operating within a vast residential area of a metropolis is likely to be confronted with quite evenly distributed customer locations. On the contrary, a vehicle operating within a rural area may rather encounter many small or medium aggregations of customer locations

[1] In order to ensure the resulting scenarios' soundness, both extremely high and extremely low candidate values for α_i are discarded and replaced by the next moderate value to be realized, i.e., a customer's request probability α_i must satisfy $0.1 \leq \alpha_i \leq 0.9$.

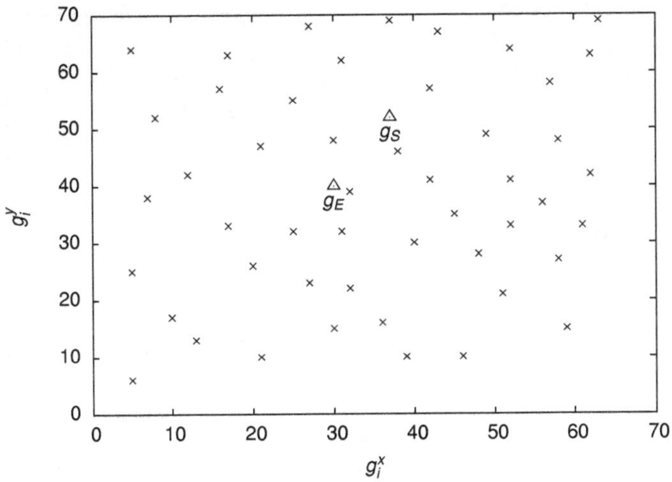

Fig. 8.1 The EV service region featuring evenly distributed customer locations.

with relatively long distances in between the aggregations. In addition, practical scenarios may yield intermediate types of distributions, comprising a mixture of aggregations and areas with evenly distributed customer locations.

In order to reflect the spectrum of distributions that may occur, the three geographies displayed in Figures 8.1, 8.2 and 8.3 are considered for the computational experiments. Each of the figures represents a service region with 49 customer locations as well as a start depot and an end depot. Figure 8.1 features a quite even distribution of customer locations within the service region. In contrast, a service region comprising five clusters of customer locations is proposed by Figure 8.2. Eventually an intermediate type of service region with semi-clustered customer locations is shown in Figure 8.3.

On each of the service regions, each customer is assigned a request probability, which is consulted whenever the corresponding customer is not selected as an early request customer. For each of the three service regions experiments are conducted with respect to three times ten different sets of early requesting customers (ERC-sets). Ten ERC-sets are generated with five, with ten and with twenty customers respectively. Every single ERC-set is generated randomly with each of the 49 customers of a region having the same probability of being selected as an element of the set.

Altogether, the experiments focus on three different service regions as well as three different fractions of early requesting customers, leading to a total of ninety different ERC-sets under consideration. For the sake of simplicity, the three service regions are referred to as CL (clustered), SC (semi-clustered) and EV (even) in the following.[2]

[2] The service regions are derived from TSP test instances present in the literature. The EV region of Fig. 8.1 emulates the `eil51`-instance as available from the TSP-library (TSPLIB, 2010). The

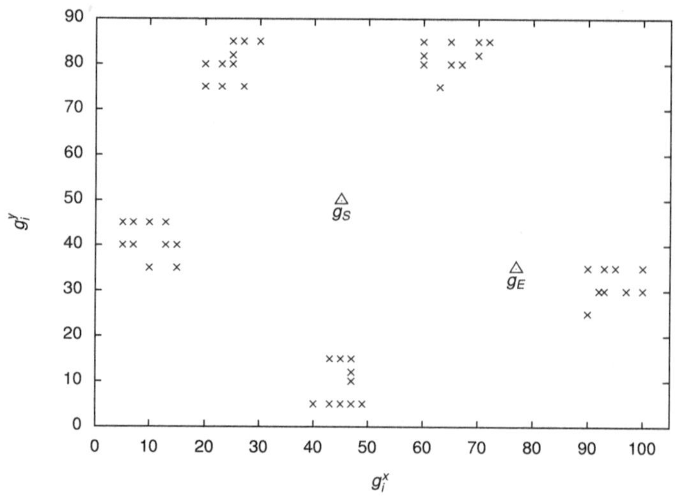

Fig. 8.2 The CL service region featuring clustered customer locations.

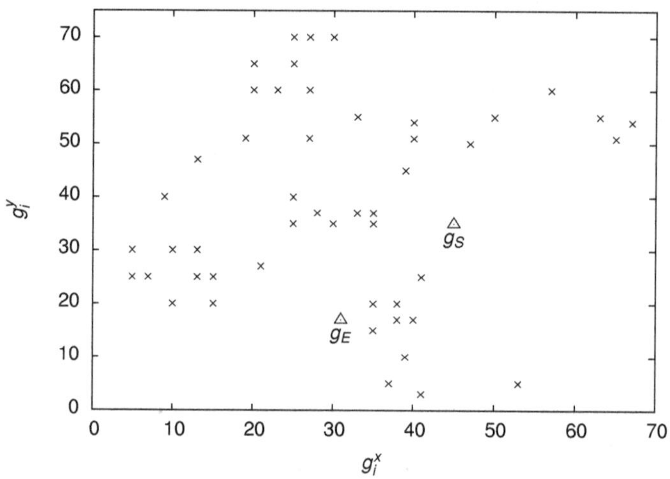

Fig. 8.3 The SC service region featuring semi-clustered customer locations.

8.1.2 Actor-Critic Configuration

The family of actor-critic methods defined in Sect. 7.2 opens up a large variety of configurations for realizing approximate anticipation. Only a subset of these

CL region and the SC region displayed in Figs. 8.2 and 8.3 rely on variants of the `c101`-instance and the `c201`-instance as provided by Solomon (1987). Every test instance proposes a total of 51 geographical locations, each of which is assigned a unique identifier id $\in [1 \ldots 51]$. For each service region, locations number 1 and number 51 are selected to represent the start depot and the end depot respectively.

configurations can be consulted within the scope of the following computational experiments. Thus, a number of presettings must be made with respect to the choices discussed in Sect. 7.2.4. Strict presettings are imposed with respect to the stepsize rules, the value estimates and the decision model structure to be consulted. Moreover, the experiments rely on the computationally more efficient exploitation state sampling.

Considering only two candidate move operations per set of confirmed customers reduces the computational burden provoked by both the learning phase of the actor-critic method and the implementation of the learned policy. For this reason the decision model structure of Eq. 7.12 is preferred to the structure of Eq. 7.8 regardless of the implied reduction of the degree of anticipation.

Further presettings are made for decision model identification. For reduction of the computational burden, the feasibility of a candidate decision is verified by means of only a single GRASP-iteration and a restricted candidate list of size one. Local search is applied merely at the very beginning of a state trajectory for generating an initial route comprising the early requesting customers. Subsequent decisions are made without local search. In order to facilitate comparability, exactly the same configuration of Procedure 11 is applied to the non-reactive approaches of Sect. 7.3.

The results of non-reactive anticipation are contrasted with results gained from the two types of value function approximation proposed in Sect. 7.2.1. Both the elementary value function approximation according to Eq. 7.5 and the more fine-grained approximation according to Eq. 7.7 are considered. Thus, initial values must be set for 350 parameters in the former case and for a total of $350 \times 49 = 17,150$ parameters in the latter case. As no specific information about a promising initial parameter setting is available, the default value of zero is used for initialization. This presetting implies that the initial policy of the actor-critic method is similar to the policy represented by the RW approach of Sect. 7.3.2.

However, each of the parameters does not only need an initial value, but also requires its own stepsize rule (cf. Sect. 5.2.2). In order to minimize the effort spent on parameter tuning, the same constant stepsize rule is preset for every parameter. As a consequence, the whole actor-critic method is parameterized by only two values. Both the stepsize value and the value of λ must be set. To this end, thirty runs of an actor-critic method using the elementary value function approximation of Eq. 7.5 have been conducted. The empirical results of these runs suggest $\lambda = 0.7$ as well as a constant stepsize value of 0.0003 to be used for the computational study.[3]

[3] The thirty runs of an actor critic method were executed for two different ten customer ERC-sets on the CL service region. On the one hand, two times five runs of Procedure 9 were conducted with stepsize values of 0.3, 0.03, 0.003, 0.0003 and 0.00003. On the other hand, the value of λ was determined by a total of two times ten runs with stepsize 0.0003 and $\lambda \in \{0, 0.1, \ldots, 0.9\}$. Of course a larger number of experiments is needed for a detailed characterization of both sensitivity and performance of an actor-critic method with respect to service region and number of early requesting customers. However, in order to keep the effort spent on parameter tuning as low as possible, such a detailed characterization is not provided at this point.

Table 8.1 Average solution quality of RW, DW and WS on the CL region with five early requesting customers.

ERC set	RW	DW	WS	Δ(DW,RW) in %	Δ(DW,WS) in %
1	11.76	12.78	6.85	8.64	86.57
2	11.18	12.24	6.78	9.52	80.70
3	12.04	12.53	5.80	4.09	116.15
4	11.24	11.46	7.05	1.99	62.60
5	10.96	10.94	6.54	-0.22	67.18
6	11.78	11.63	6.66	-1.32	74.59
7	11.07	11.27	6.74	1.81	67.18
8	11.54	11.68	6.64	1.18	75.84
9	11.73	13.13	6.61	11.90	98.55
10	11.48	12.97	6.40	13.00	102.63
mean	**11.48**	**12.06**	**6.61**	**5.06**	**83.20**
STD	**0.35**	**0.77**	**0.33**	**5.24**	**17.58**

8.2 Non-reactive Anticipation

The performance of non-reactive anticipation for dynamic routing of a service vehicle is analyzed in the following. Section 8.2.1 provides a comparison of the conventional approaches introduced in Sect. 7.3.2. Subsequently, Sect. 8.2.2 puts the focus on the probabilistic approaches of Sect. 7.3.1 and contrasts their performance with the results developed in Sect. 8.2.1.

8.2.1 Conventional Approaches

The three competing conventional approaches WS, RW and DW act upon different principles. WS seeks for advantages by allocating the maximum amount of waiting time at the start depot. In contrast, both RW and DW strive for a balanced allocation of the available waiting time among the confirmed customers. RW tends to be more relaxed than DW, because its stochastic nature sometimes provokes early moves where DW forces the vehicle to wait deterministically. In theory, RW may also lead to longer waiting times than effected by DW. However, such a situation will only occur if the total amount of waiting time DW assigns to a single customer is reduced to a minimum. Thus, in most of the cases, RW "moves faster" than DW.

Table 8.1 illustrates the performance of RW, DW and WS on the CL region with five early requesting customers. Ten different ERC-sets are considered. For each set the quality of each of the three approaches is determined by averaging the number of confirmed late requests over 250 state trajectories. Given a specific trajectory $k \in [1 \ldots 250]$, each customer i implements exactly the same behavior for each of the ten settings, unless i is selected as an element of the current ERC-set. Subject to these conditions, RW yields an average of 11.76 served late requests over the 250

Table 8.2 Average solution quality of RW, DW and WS on the CL region with ten early requesting customers.

ERC set	RW	DW	WS	Δ(DW,RW) in %	Δ(DW,WS) in %
1	9.49	9.50	4.73	0.13	101.02
2	8.75	8.82	4.92	0.75	79.08
3	10.80	11.23	4.74	3.96	137.16
4	9.09	9.67	5.16	6.38	87.36
5	10.18	10.84	5.47	6.44	98.02
6	9.88	10.10	4.03	2.14	150.65
7	9.22	9.89	5.27	7.25	87.70
8	10.05	10.43	4.44	3.74	134.86
9	9.90	10.52	5.00	6.18	110.15
10	10.32	9.93	3.80	-3.76	161.54
mean	**9.77**	**10.09**	**4.76**	**3.32**	**114.76**
STD	**0.63**	**0.70**	**0.54**	**3.52**	**29.11**

trajectories for ERC-set number one. For the same set, WS only leads to an average of 6.85 confirmations, whereas DW performs best by showing 12.78 confirmations on average. The difference Δ(DW,RW) of the averages of DW and RW corresponds to 8.64% of the solution quality of RW, i.e., DW exceeds the quality of RW by 8.64%. In the same sense, DW is 86.57% better than WS, if ERC-set one is in the CL region.[4]

DW performs best for eight out of the ten ERC-sets considered. The remaining two sets are dominated by RW. In any case both DW and RW are significantly superior to WS. The performance gap between DW and WS is illustrated by the rightmost column of Table 8.1 showing DW to exceed WS by at least 62.6 % and by 83.2% on average. In contrast, DW exceeds RW by 5.06% on average, with the maximum surplus of DW being 13% and the maximum deficit of DW being -1.32%. However, DW exceeds RW by more than 2% on half of the ERC-sets

Increasing the number of early requesting customers from five to ten entails two major impacts on the relative behavior of RW, DW and WS. Merely at first glance the behavior seems quite similar to the behavior observed with five early requesting customers. As shown in Table 8.2, DW again yields the largest sample mean over the ten ERC-sets and DW again almost always outperforms both RW and WS. On the contrary, the results of DW and RW are more similar with ten early requesting customers than they are with five. Now DW exceeds RW on average by only 3.32% instead of 5.06%. At the same time the sample standard deviation of the relative performance of the two reduces from 5.24% down to 3.52%. Moreover, with ten early requesting customers the maximum surplus of DW over RW is only roughly half of the maximum surplus with five early requests.

[4] For the purpose of presentation, the values in Table 8.1 as well as in the following tables in this chapter were rounded to two positions after the decimal point. Thus, the values within the two rightmost columns may not precisely match the data of the other columns.

Table 8.3 Average solution quality of RW, DW and WS on the CL region and twenty early requesting customers.

ERC set	RW	DW	WS	$\Delta(\text{RW},\text{DW})$ in %	$\Delta(\text{RW},\text{WS})$ in %
1	8.43	8.50	2.67	-0.85	215.89
2	6.72	6.62	2.92	1.45	130.00
3	6.21	6.04	2.14	2.85	190.09
4	7.35	7.31	2.81	0.49	161.68
5	8.58	8.56	3.08	0.14	178.08
6	7.90	7.73	2.78	2.17	184.44
7	6.19	6.05	2.06	2.25	200.39
8	6.54	6.58	1.76	-0.55	271.82
9	8.16	8.12	2.90	0.59	181.13
10	7.24	7.09	2.46	2.14	193.83
mean	**7.33**	**7.26**	**2.56**	**1.07**	**190.74**
STD	**0.90**	**0.95**	**0.44**	**1.28**	**36.72**

Table 8.4 Average solution quality of DW, RW and WS on the CL region.

CL region		RW	DW	WS	$\Delta(\text{best},\text{2nd})$ in %	$\Delta(\text{best},\text{WS})$ in %
5 ERC	mean	11.48	**12.06**	6.61	5.06	83.20
	STD	0.35	0.77	0.33	5.24	17.58
10 ERC	mean	9.77	**10.09**	4.76	3.32	114.76
	STD	0.63	0.70	0.54	3.52	29.11
20 ERC	mean	**7.33**	7.26	2.56	1.07	190.74
	STD	0.90	0.95	0.44	1.28	36.72

The second major impact of an increased number of early requesting customers is related to the quality of WS. Compared to both RW and DW, WS performs much worse than before. The gap between DW and WS increases tremendously and now shows an average of 114.5% with the maximum surplus of DW over WS being 161.54%.

The trends observed in the course of increasing the number of early requesting customers from five to ten are substantiated by Table 8.3. This table shows the results of RW, DW and WS for ten randomly generated sets of twenty early requesting customers. Obviously the further increase leads to both a further harmonization of DW and RW and a further increase of the performance gaps between WS and DW as well as WS and RW. The latter exceeds WS by at most 271.82% with an average surplus of 190.74%.

Yet, the most striking change with respect to the previous settings is the fact that in the twenty customer case, DW is no longer best. RW outperforms DW on eight out of ten ERC-sets and exceeds it by 1.07% on average. In the best case, RW yields an advantage of 2.85%, wheras in the worst case its deficit is at a level of less than one percent.

A summary of the conventional non-reactive approaches' performance on the CL region is provided in Table 8.4. The major conclusions are:

- A balanced allocation of waiting time as pursued by RW and DW is always superior to allocating the maximum amount of waiting time at the start depot. In particular, WS is outperfomed by both RW and DW on each of the ERC-sets considered.
- The more early requesting customers are present, the stronger is the superiority of a balanced allocation in contrast to waiting at the start.
- The more early requesting customers are present, the more similar perform DW and RW. Both the average relative difference of the two and the corresponding sample standard deviations decrease if the number of early requesting customers is increased.
- DW outperforms RW on average for both five and ten early requesting customers. In the case of twenty early erquesting customers RW outperforms DW. In any case, the best approach does not perform best on every ERC-set.

More insights may be gained by contrasting these conclusions with the intended behavior of the approaches. As discussed in the introduction of this section, RW tends to move faster than DW, which explains the superiority of DW with respect to five and ten early requests. If for example at a certain time only one of the customers within an agglomeration issued a request and received a confirmation, then serving this customer and moving on to the next agglomeration at once is likely to be misleading. As soon as more customers from the first agglomeration request for service, the vehicle may either return or not. In the former case, another long distance trip between two agglomerations is required, whereas in the latter case, the new requesting customers must be rejected. Both cases are not preferable and may be avoided by the longer waiting times of DW. Obviously, this finding is of particular relevance at the very beginning of a shift with only a few early requesting customers. The less early requesting customers are present, the more likely are sparse confirmations within the agglomerations at the beginning.

In the presence of twenty early requesting customers DW moves too slow. In this situation it is likely that already at the beginning of the shift a relatively high number of confirmations is present in every agglomeration. Thus, the vehicle may not need as much waiting time as before in order to remain long enough within one agglomeration, i.e., within one agglomeration serving customers compensates waiting operations while reducing the number of locations to be visited in the future. In addition, the superiority of RW is supported by the tight schedule imposed by the large number of early requests. Both the mere number of locations to be visited and the many long distances to be travelled for covering the agglomerations effect a quite small amount of slack already at the beginning of the shift. In this case, DW's relatively slow moves are disadvantageous, because moving slow leads to a growing number of unserved confirmations. These imply a rapid decrease of the degrees of freedom with respect to future decisions. Exactly the same reasoning may explain both the bad performance of WS and the worsening of WS if the number of early requests increases.

Table 8.5 Average solution quality of DW, RW and WS on the SC region.

SC region		RW	DW	WS	Δ(best, 2nd) in %	Δ(best, WS) in %
5 ERC	mean	**15.19**	14.91	10.54	1.94	44.30
	STD	0.45	0.50	0.52	0.83	3.47
10 ERC	mean	**13.06**	12.92	8.21	1.15	59.60
	STD	0.54	0.60	0.56	0.94	10.12
20 ERC	mean	8.74	**8.77**	4.52	0.34	95.35
	STD	0.86	0.82	0.60	1.03	14.02

Table 8.6 Average solution quality of DW, RW and WS on the EV region.

EV region		RW	DW	WS	Δ(best, 2nd) in %	Δ(best, WS) in %
5 ERC	mean	**14.39**	13.51	9.28	6.55	55.19
	STD	0.40	0.45	0.38	1.07	5.07
10 ERC	mean	**11.75**	11.31	6.76	3.96	74.43
	STD	0.40	0.38	0.43	1.34	9.09
20 ERC	mean	7.08	**7.39**	2.85	4.53	164.00
	STD	0.83	0.74	0.54	2.09	33.94

Finally, the properties of DW explain the increasing assimiliation of DW and RW subject to a growing number of early requests. More early requests imply a reduction of the amount of waiting time DW assigns to a customer within the current route. As a consequence, the time between two moves is reduced and the possible advantage effected by an early move of RW vanishes. The tendency towards assimilation can also be observed if the number of early requests is increased on both the EV region and the SC region.

An overall summary of the the three conventional approaches' performances on the SC region and the EV region is provided in Table 8.6 and Table 8.5 respectively. The tables show that the conclusions drawn with respect to WS on the CL region hold independent of the service region considered. WS always performs worst and always entails a performance decrease if the number of early requests grows. Nevertheless, Tables 8.6 and 8.5 allow for additional conclusions:

- Compared to the experiments on the CL region, the performances of RW and DW are reversed for both the SC region and the EV region. The two latter regions show RW to be better than DW in the case of five early requests and in the case of ten early requests. In contrast, DW is superior to RW on both regions in the case of twenty early requests.
- Given a specific number of early requests, the advantage of RW and DW over WS diminishes if the service region becomes more even. For instance, in the case of ten early requests, WS is exceeded by 114.76% on the CL region as opposed to 74.43% on the SC region and only 59.60% on the EV region.

The reversed performance of RW and DW for both five and ten early requests may be explained by the significant change of the service regions' structure. In contrast to the CL region, both the EV region and the SC region are less restrictive, because they provide customer locations in between two subregions that are a long way away from each other. Therefore, expensive long distance return trips as provoked by RW on the CL region are less likely to occur. As a consequence, the extremely long waiting times assigned by DW in the presence of only a few confirmations tend to be a waste of time.

However, the relatively early moves provoked by RW are not always recommended. In the presence of twenty early requests, RW moves too fast, because it cannot compensate its lack of waiting time by short distance moves as it can within an agglomeration of the CL region. At the same time DW improves, because it assigns a smaller amount of waiting time to each customer than it does in the presence of less early requests.

In summary, the analysis shows that the adequateness of RW and DW heavily depends on the problem structure in terms of both the number of early requests and the type of service region. Similarly, the deficit of WS is strongly influenced by the problem structure. The relative performance of WS improves if the problem structure gets more relaxed, i.e., WS shows its best relative performance if the customer locations are distributed evenly and the number of early requests is low. In contrast, the superiority of a balanced allocation of waiting time is maximal in case of a high number of early requests in the CL region. An explanation may be that not moving in the face of a strict problem structure leads to a quite rapid decrease of the degrees of freedom with respect to future decisions.

8.2.2 Probabilistic Approaches

In contrast to the methods of the previous section, the three probabilistic approaches PC, LW and CW each consult the customers' request probabilities for making a decision on whether to wait or not. Both LW and CW capture potential customers by means of their center of gravity and determine one single dedicated waiting position for the current decision. LW identifies the position allowing for latest departure whereas the position derived by CW aims at slack maximization. PC does not derive a dedicated waiting position per decision. Instead it makes decisions by considering both the number of actual confirmations and the number of presumed customers as a criterion on whether to wait or not at the current location.

Table 8.7 shows the results of PC, LW and CW on the CL region subject to the same ten sets of five early requesting customers as considered in Table 8.1. Among the probabilistic approaches, LW performs best on each of the sets. CW generates the second best result in nine out of ten cases, whereas PC is second best only once. Though this imposes a quite strict hierarchy of the probabilistic approaches, the differences between PC, LW and CW are much smaller than the differences between the conventional approaches emerging from Table 8.1. The average difference

Table 8.7 Average solution quality of PC, LW and CW on the CL region with five early requesting customers.

ERC set	PC	LW	CW	$\Delta(LW,DW)$ in %	$\Delta(LW,PC)$ in %
1	11.28	11.91	11.54	-6.79	5.53
2	10.85	11.29	11.12	-7.81	4.06
3	11.10	11.58	11.39	-7.57	4.32
4	10.95	11.20	11.16	-2.23	2.34
5	9.66	11.07	10.86	1.17	14.58
6	10.96	11.85	11.83	1.89	8.06
7	10.75	11.20	10.87	-0.64	4.13
8	11.02	11.57	11.26	-0.89	5.05
9	11.00	11.31	11.09	-13.86	2.80
10	11.38	11.74	9.45	-9.50	3.09
mean	**10.90**	**11.47**	**11.06**	**-4.62**	**5.4**
STD	**0.47**	**0.30**	**0.64**	**5.21**	**3.61**

Table 8.8 Average solution quality of PC, LW and CW on the CL region with ten early requesting customers.

ERC set	PC	LW	CW	$\Delta(LW,DW)$ in %	$\Delta(LW,PC)$ in %
1	8.22	9.70	9.36	2.06	18.06
2	8.28	8.54	8.35	-3.20	3.04
3	8.18	11.50	10.07	2.42	40.70
4	7.57	9.03	8.83	-6.58	19.34
5	9.48	10.29	9.96	-5.06	8.48
6	6.96	9.93	9.45	-1.66	42.64
7	8.12	9.37	8.15	-5.26	15.31
8	9.12	9.97	9.73	-4.37	9.29
9	7.89	9.80	9.54	-6.77	24.29
10	9.75	10.80	10.12	8.74	10.75
mean	**8.36**	**9.89**	**9.36**	**-1.97**	**19.19**
STD	**0.86**	**0.85**	**0.70**	**4.97**	**13.33**

between the best (LW) and the worst (PC) of the probabilistic approaches is at approximately the same level as the corresponding average difference between the best (DW) and the second best (RW) of the conventional approaches.

However, the probabilistic information about upcoming customer requests can on average not be converted into the best number of confirmations. On average, LW is 4.62% worse than DW, which means that the deficit of LW with respect to DW is similar to the deficit of RW with respect to DW. DW outperforms LW on eight out of ten ERC-sets with the maximum deficit of LW being 13.86% as opposed to a maximum surplus of LW of only 1.89%.

LW still is the best of the probabilistic approaches in the CL region if the number of early requests is increased to ten. Table 8.8 shows that LW outperforms both CW and PC on each of the same ten ERC-sets as considered in Table 8.2. LW now

Table 8.9 Average solution quality of PC, LW and CW on the CL region with twenty early requesting customers.

ERC set	PC	LW	CW	Δ(LW,RW) in %	Δ(LW,PC) in %
1	7.86	8.66	8.22	2.71	10.07
2	5.87	6.58	6.19	-1.97	12.13
3	5.55	6.12	5.90	-1.61	10.01
4	6.09	7.70	6.96	4.79	26.40
5	6.83	8.76	8.18	2.10	28.24
6	6.82	7.80	7.53	-1.17	14.50
7	5.30	6.45	5.78	4.20	21.75
8	5.53	6.64	6.28	1.41	19.96
9	6.55	8.03	7.78	-1.67	22.60
10	5.98	7.15	6.94	-1.22	19.60
mean	**6.24**	**7.39**	**6.97**	**0.76**	**18.52**
STD	**0.78**	**0.94**	**0.92**	**2.60**	**6.56**

Table 8.10 Average solution quality of PC, LW and CW on the CL region.

CL region		PC	LW	CW	Δ(LW,conv*) in %	Δ(LW,PC) in %
5 ERC	mean	10.90	**11.47**	11.06	-4.62	5.40
	STD	0.47	0.30	0.64	5.21	3.61
10 ERC	mean	8.36	**9.89**	9.36	-1.97	19.19
	STD	0.86	0.85	0.70	4.97	13.33
20 ERC	mean	6.24	**7.39**	6.97	0.76	18.52
	STD	0.78	0.94	0.92	2.60	6.56

exceeds PC by 19.19% on average, whereas the difference between LW and CW is approximately the same as in the case of five early requests. LW also improves its performance with respect to DW. However, its average quality still is approximately 2% less than the quality produced by DW. In seven out of ten cases DW outperforms LW, with the latter being at most 6.77% worse than the former. Anyway, in one case LW is superior by 8.74%, which indicates that a balanced distribution of the waiting time is not a general winning strategy in the CL region.

A further endorsement of this hypothesis is provided by Table 8.9 showing the results of the probabilistic approaches in the presence of twenty early requests in the CL region. Subject to this setting, the performance gaps between PC, LW and CW remain at approximately the same level as subject to ten early requests. However, LW is now on average even slightly superior to the best among the conventional approaches (RW). Nevertheless, the situation is not clear-cut as LW outperforms RW on only half of the ERC-sets considered.

A summary of the overall performance of the three probabilistic approaches on the CL region is provided by Table 8.10. The two most prominent findings are:

Table 8.11 Average solution quality of PC, LW and CW on the SC region.

SC region		PC	LW	CW	Δ(LW,conv*) in %	Δ(LW,PC) in %
5 ERC	mean	13.55	**15.41**	15.33	1.46	13.96
	STD	0.72	0.48	0.45	0.49	4.93
10 ERC	mean	11.90	**13.32**	13.21	2.00	12.06
	STD	0.57	0.56	0.56	0.62	3.41
20 ERC	mean	7.95	**9.02**	8.80	2.84	13.31
	STD	0.64	0.89	0.84	0.95	5.20

- LW clearly dominates PC and CW independent of the number of early requests. PC is by far worst for most of the ERC-sets. The relatively good average performance of PC in the case of five early requests may indicate that PC copes quite well with scarce confirmations within an agglomeration.
- LW is dominated by the best of the conventional approaches (denoted as "conv*" in Table 8.10) in case of both five and ten early requests. LW still is outperformed on five out of ten ERC-sets in the case of twenty early requests. However, in this case LW is on average slightly better than the best of the conventional approaches.

More general conclusions may be drawn by additionally taking into account Tables 8.11 and 8.12. These tables illustrate the performances of the probabilistic approaches on the SC region and on the EV region respectively. The additional results lead to the following overall summary:

- The hierarchy of LW, CW and PC is the same on each of the service regions. Among the probabilistic approaches, LW performs on average best in every service region independent of the number of early requests.
- If the service region becomes more even, the performance gaps between the three approaches tend to reduce. In the CL region, LW outperforms PC by approximately 20% on average for both ten and twenty early requests. In contrast, LW is ahead of PC by 12% to 14% if the customer locations are semi-clustered. Finally, the differences between the two reduce to less than 10% in the EV region. At the same time the gap between LW and CW decreases from up to 6% in the CL region to at most 1.9% in the EV region.
- The less early requests are present in a service region, the more similar perform LW and CW. The performances of the two fully assimilate in the presence of five early requests in the EV region.
- If the number of early requests in a service region grows, the performance of LW tends to improve with respect to the performance of the corresponding best conventional approach.
- Given a specific number of early requests, the performance of LW increases with respect to the best conventional approach if the customer locations are distributed more evenly. In the CL region, five early requests entail a 4.62% deficit of LW with respect to DW. In constrast, LW is ahead of RW by 1.46% in the SC region

Table 8.12 Average solution quality of PC, LW and CW on the EV region.

EV region		PC	LW	CW	Δ(LW, conv*) in %	Δ(LW, PC) in %
5 ERC	mean	13.39	**14.61**	**14.61**	1.52	9.18
	STD	0.45	0.40	0.42	0.52	3.97
10 ERC	mean	11.11	**12.15**	12.03	3.37	9.52
	STD	0.48	0.36	0.38	0.93	6.75
20 ERC	mean	7.05	**7.64**	7.50	3.30	8.34
	STD	0.70	0.84	0.75	2.40	4.29

and by 1.52% in the EV region. In the more favorable case of twenty early requests LW entails an average surplus of 0.76% in the CL region as opposed to advances of 2.84% and 3.3% in the SC region and the EV region respectively.

Clearly, the LW principle of waiting at the position that allows for latest departure is superior to the principles of CW and PC. Although CW outperforms LW for some of the ERC-sets in the EV region, the experiments suggest CW's principle of waiting in hope of slack maximization to be less successful in general. Similarly, the PC principle of considering a presumed customer to be as important as an actual confirmation must be discarded. The fact that PC stays far behind LW suggests that it puts too much emphasis on the future in most of the cases.

As exemplified by the comparison of LW with the respective best conventional approach, the adequateness of the probabilisitc approaches depends to a large extent on the problem structure. The increasing adequateness of LW in the presence of many early requests may be reduced to the fact that a higher number of elements within the current route enables more precision in the selection of the dedicated waiting position. At the same time, the declining precision induced by less early requests may trigger the observed assimiliation of the performances of LW and CW.

Just as an increasing number of early requests, a more even distribution of customer locations tends to provoke more precision. In case of an even distribution, the center of gravity is more sensitive with respect to request probabilities. Thus it is likely to capture the location of upcoming requests more precisely than in the case of overlapping effects of probabilities and a strong variation of distances among the customer locations.

As a general conclusion it may be stated that the use of probabilistic information in terms of LW leads to advantages over the conventional approaches in most of the cases. LW is generally recommended in both the EV region and the SC region. However, the general recommendation changes in favor of the conventional approaches if the CL region is considered. Apart from LW's significant deficits in case of both five and ten early requests, the twenty early request case turns out to be quite ambivalent. A definite recommendation may hardly be formulated, because LW and RW are at approximately the same level on average with considerable differences on the single ERC-sets.

8.3 Elementary Value Function Approximation

The insights gained from the preceding section are now contrasted with approximate anticipation in terms of the elementary value function approximation derived from Hypothesis 1 in Sect. 7.2.1. As this approach relies on both the number of presumed customers and Hypothesis 1, it is referred to as "PC-1" in the following. Section 8.3.1 illustrates the behavior of PC-1 by means of representative examples. Then, Sect. 8.3.2 focuses on comparing the results achieved by the elementary value function approximation to the results of the non-reactive approaches.

8.3.1 Solution Properties

Consulting value functions that are derived from the information structure of Eq. 7.5 implies a total of $\mathscr{T} = 350$ parameters to be determined. A single parameter r_t reflects the importance of the presumed number of customers in making a decision at time t. Note that such parameters are also present in both the PC approach and the RW approach. PC assumes a presumed customer to be as important as a customer that currently receives a confirmation, i.e., $\forall t : r_t = 1$. In contrast, RW does not take into account presumed customers but only focuses on the customers actually requesting for service, i.e., $\forall t : r_t = 0$. However, both of these approaches represent a non-reactive type of anticipation. That is, both of them assume parameter values that are not derived from the behavior of the remaining decision process.

PC-1 produces parameter values in response to the behavior of the remaining decision process. Figure 8.4 displays such values for an exemplary set of five early requests in the EV region. The values are the result of 500,000 iterations of an actor-critic method operating subject to the presettings discussed in Sect. 8.1.2.

Evidently, the actor-critic method derives values that are significantly different from both 0 and 1 for $t \in [0 \dots 339]$. Although the curve is rugged at the beginning, the parameter values may be described as being quite close to 0.2 for $t \in [0 \dots 60]$. From $t = 60$ on, the values are subject to a roughly linear increase up to approximately 0.5 at $t = 216$. Subsequently, there is a quite smooth decrease down to 0 at $t = 340$. The resulting pile-like shape of the curve indicates a continuous increase of the importance of the number of presumed customers until $t = 216$ and a relatively rapid decrease of its importance afterwards. An explanation of this shape might be that a compensation of decisions which are potentially disadvantageous with respect to the future is more difficult, if such decisions are made at later decision times. Thus, an increased focus on the presumed customers is necessary as time passes by. Moreover, the decline beyond $t = 216$ might indicate that a diminishing amount of slack is increasingly restrictive with respect to possible future confirmations, i.e., with not much time left it is better to increase the focus on immediate confirmations instead of taking into account uncertain candidates in the future.

Additional analysis is required in order to understand the detailed characteristics of the curve. At a more detailed level, it can be recognized that the ruggedness of

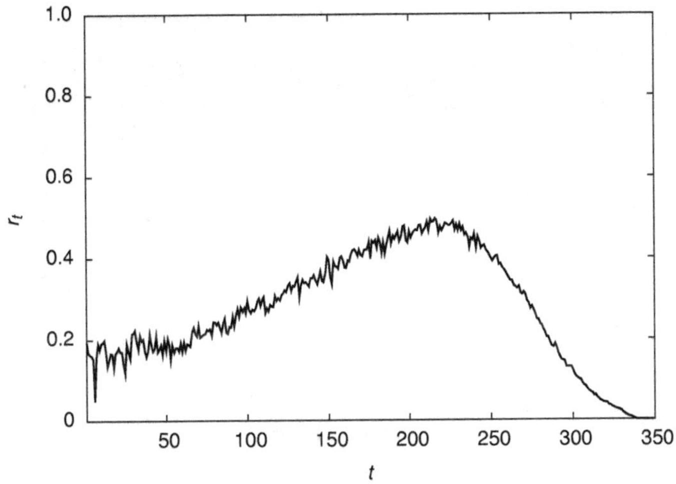

Fig. 8.4 The values of the parameters r_t of PC-1 after 500,000 iterations.

the curve reduces as t increases. On the one hand this phenomenon may be due to the fact that an earlier point in time t is in general subject to a higher variation of the sample estimates used for updates at t. On the other hand the relatively strong ruggedness within the very early points in time also results from a lack of updates. Consider the striking low point at $t = 6$ as an example. While almost every parameter value is based on thousands of updates, r_6 only received two updates throughout the first 500,000 iterations. Thus, the current value of roughly 0.05 is likely to be an imprecise estimate of the optimal parameter value at $t = 6$. In general, the fact that a point in time t receives only a very small number of updates means that the applied policies do not consider t as a decision time very often. The experiments show that such points occur exclusively at the very beginning of the time horizon, because throughout the iterations the very beginning proves to be dominated by quite similar vehicle moves. For example, in most of the cases, the vehicle leaves the start depot within the first five time units and moves to one of the early requesting customers, which prevents decisions at $t = 6$ in any case.

Figures 8.5, 8.6 and 8.7 illustrate the evolutions of both parameter value and number of updates for r_{100}, r_{200} and r_{300} respectively. The value of r_{100} reaches a quite stable level of approximately 0.26 already after 10,000 iterations and less than 800 updates. In contrast, r_{200} needs about 70,000 iterations and 5,000 updates until it remains at a level of roughly 0.47. The value of r_{300} takes by far the longest time of the three for reaching a final level. It climbs up to about 0.115 after roughly 700,000 iterations and nearly 60,000 updates.

As the number of updates grows at approximately the same ratio for each of the three parameters, the different numbers of iterations needed for stabilization, suggest different speeds of convergence. The parameter values at later decision times seem to converge much slower than the values at earlier times. This suggests earlier decisions to exert a quite heavy influence on the sample estimates of later decision

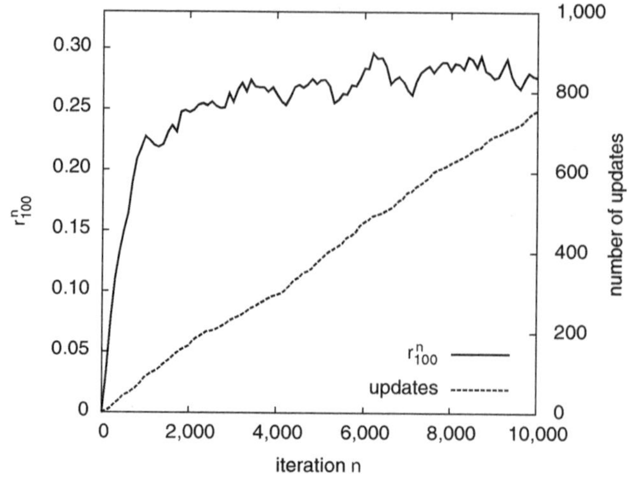

Fig. 8.5 Evolution of both value and updates of parameter r_{100} over the first 10,000 iterations.

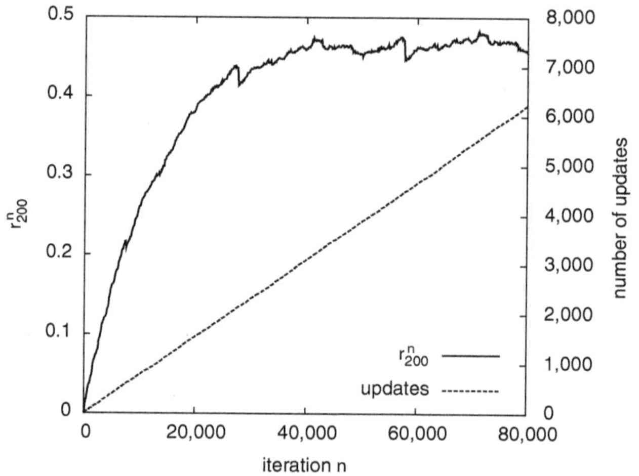

Fig. 8.6 Evolution of both value and updates of parameter r_{200} over the first 80,000 iterations.

times while on the contrary, the contribution of later decisions to the sample estimates of the preceding decision times is relatively low. Thus, convergence of the earlier parameter values is a precondition of convergence at later decision times.

In addition, the parameter curves of earlier decision times show more ruggedness than the corresponding curves of later decision times. As already discussed with respect to Fig. 8.4, the increased ruggedness within the evolution of earlier parameter values may be attributed to a higher variation of the sample estimates subject to a fixed policy, i.e., the stochastic influence of the customer behavior is more intense with respect to earlier decision times than with respect to later decision times.

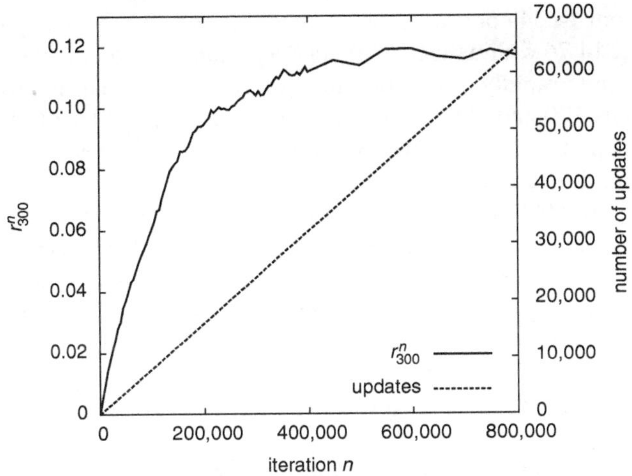

Fig. 8.7 Evolution of both value and updates of parameter r_{300} over the first 800,000 iterations.

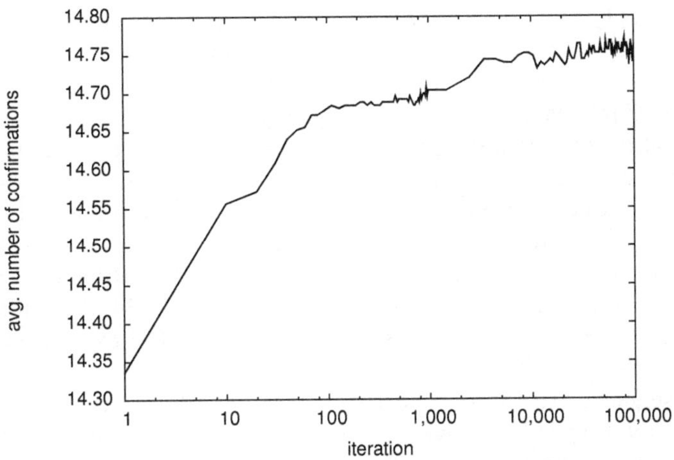

Fig. 8.8 Exemplary evolution of the solution quality of PC-1 subject to five early requests in the EV region.

More information about the parameters can be extracted from Fig. 8.8. This figure shows the evolution of the solution quality of PC-1 over the first 100,000 iterations. It displays the quality of every 10th parameter set between iteration one and iteration one hundred, the quality of every 20th set between iterations one hundred and one thousand as well as the quality of every 1,000th parameter set occuring from iteration one thousand on. Each of these sets is applied to the same 250 test-trajectories and evaluated by the average number of confirmed late requests from these trajectories.

Starting from 14.34 confirmations on average, the solution quality climbs up to approximately 14.76 within the first 100,000 iterations. Up to iteration 100 the quality increases quite rapidly before turning into a phase of small improvements between iterations 100 and 800. Thereafter, a couple of jumps lead to an average of 14.74 confirmations at iteration 10,000, from which on a number of minor improvements occur until iteration 100,000.

At iteration 100,000 the course of the parameter values over the decision times is only slightly more rugged than the course at iteration 500,000 which is shown in Fig. 8.4. In particular, the major effects of iterations 100,000 to 500,000 turn out to be a slight reduction of the overall ruggedness as well as a better alignment of the values at decision times lacking of updates. For instance, the above discussed parameter at decision time $t = 6$ did not receive any update up to iteration 100,000, while its value is raised from 0 to 0.05 by means of two updates up to iteration 500,000.

In addition, also Figs. 8.5, 8.6 and 8.7 indicate that typically only minor changes of the parameter values occur beyond iteration 100,000. Both r_{100} and r_{200} stabilize much earlier than iteration 100,000. Although r_{300} does not stabilize yet by this iteration it is only about 0.05 away from its final level. The fact that the parameters of very late decision times are not yet on a stable level while at the same time the improvements of the solution quality become smaller, suggests that earlier decisions exert a higher influence on the solution quality than later decisions. This seems reasonable also from an intuitive point of view, because earlier decisions determine to a large extent the degrees of freedom of later decisions.

Evaluating the best parameter set according to Fig. 8.8 by means of the same 250 trajectories as considered in Sect. 8.2 for evaluation of the non-reactive approaches, yields an average solution quality of 14.78. In contrast, PC leads to only 13.55 confirmations on average and RW results into an average of 14.34. Thus, for the given set of five early requesting customers in the EV region, PC-1 exceeds PC by about 9% and outperforms RW by approximately 3%.

A representative example of the behavior of PC-1 in the presence of an increased number of early requests is provided in Fig. 8.9. The figures illustrate both the parameter values and the evolution of the solution quality for a specific set of twenty early requesting customers in the EV region. As in the five early requesting customers example, the parameter-curve shows a pile-like shape and turns out to be more rugged at the beginning than at the end. However, two major differences between Fig. 8.9a and Fig. 8.4 can be observed. First of all, the initial phase of relatively high ruggedness is prolonged in the case of twenty early requests. The experiments show that a higher number of early requests leads to the occuring policies making more similar moves in the early stage of the time horizon. This is reflected by the fact that decision times receiving a relatively low number of updates are present up to $t = 70$ in Fig. 8.9a in contrast to only up to $t = 40$ in Fig. 8.4.

In addition to the prolonged phase of ruggedness, Fig. 8.9a exemplifies that a higher number of early requesting customers tends to provoke a lower maximum level of the parameter values. The experiments show the average peak value of the parameter curves for twenty early requests to be about 0.15 below the average peak

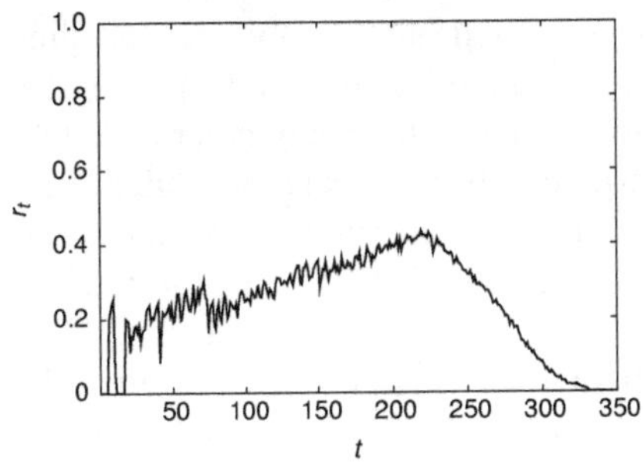

Fig. 8.9a Parameter values of PC-1 after 500,000 iterations subject to twenty early requests in the EV region.

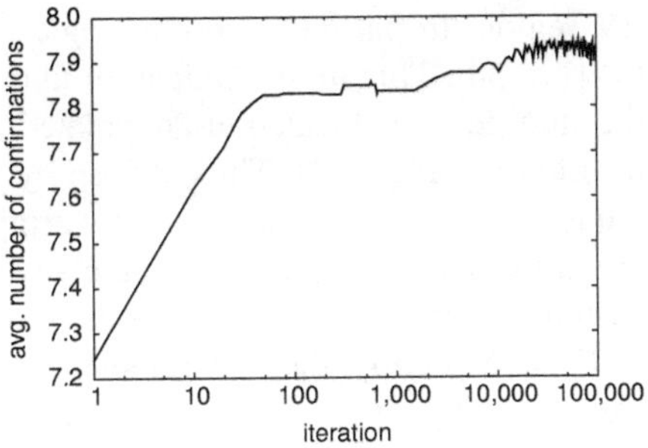

Fig. 8.9b Evolution of the solution quality of PC-1 subject to twenty early requests in the EV region.

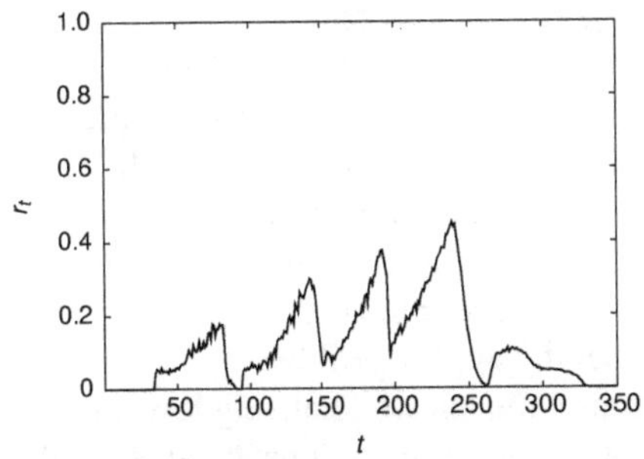

Fig. 8.10a Parameter values of PC-1 after 500,000 iterations subject to twenty early requests in the CL region.

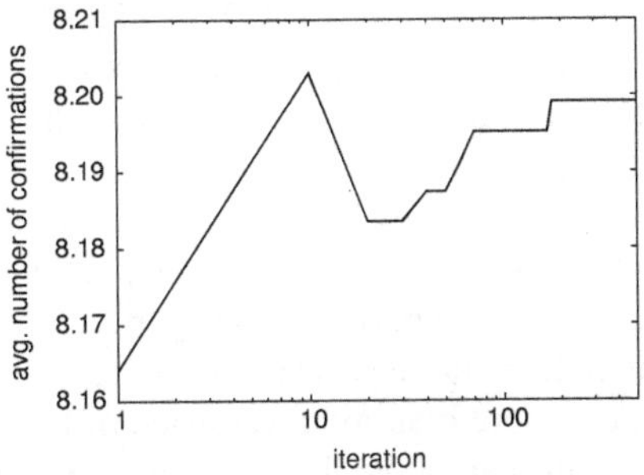

Fig. 8.10b Evolution of the solution quality of PC-1 subject to twenty early requests in the CL region.

value of the curves resulting from five early requests. This means that the reduced flexibility of vehicle movements induced by a higher number of early requests results into an increased focus on short-term confirmations.

Figure 8.9b illustrates the evolution of the solution quality for the considered set of twenty early requests in the EV region. Starting from 7.24 confirmations on average, the quality of the parameters improves up to an average of approximately 7.9. Again, the best of the displayed parameter sets is applied to the same 250 trajectories as used for evaluation purposes in Sect. 8.2. This yields an improvement of 11.87% with respect to the quality of PC as well as an improvement of 9.83% with respect to the quality of RW on the same set of early requesting customers.

The effects provoked by a change of the underlying service region are illustrated in Fig. 8.10. Both the course of the parameters over time and the evolution of the solution quality are shown for a representative set of twenty early requests in the CL region. Again the ruggedness of the parameter curve (Fig. 8.10a) decreases throughout the time horizon and again a lack of updates occurs among the very early decision times. The latter phenomenon is much more distinctive than it was with the

EV region. In particular, not a single update is carried out for any decision time $t \in [1 \ldots 34]$. This immense lack of updates is due to the fact that in the CL region, the start depot is located at an isolated position far away from any other customer location (cf. Fig. 8.2). Thus, driven by the large number of twenty sheduled early requests, none of the policies appearing throughout the iterations considers waiting at the start. Instead the vehicle always makes a long distance move to the first customer location, which causes the next decision to occur not earlier than $t = 35$.

Quite evidently, the general shape of the parameter curve differs from the pile-like shape typically observed in case of the EV region. In contrast, the CL region leads to a sawtooth-like shape with growing peak values up to $t = 240$ and a smaller additional peak at $t = 280$. The curve may be divided into five subsequent parts comprising a prominent peak value each. After the aforementioned jump from the start depot to the first customer location, the beginning of a new part may be identified at $t = 35$, at $t = 95$, at $t = 151$ as well as at $t = 197$ and at $t = 255$.

An explanation of this structure may be derived from the fact that the considered ERC-set makes the vehicle visit each of the five customer agglomerations present in the CL region. In this context, the high number of early requests and the long distances between different agglomerations impose a quite tight schedule not allowing the vehicle to return to a customer agglomeration already visited. Consequently, each of the five parts emerging from the parameter curve reflects the time interval at which the vehicle typically remains within a specific agglomeration.

At a more detailed level, each of the first four parts shows relatively low parameter values at the beginning as well as a quite steep ascent up to a peak value. Right after the peak, the value quickly drops back to a low level marking the beginning of the next part. Thus, within each of the first four agglomerations visited, early decisions are made without paying a lot of attention to presumed customers. Instead, the presumed customers are becoming more important throughout the period of time the vehicle remains within a specific agglomeration. The abrupt relapses at the end of the parts suggest that the presumed customers are of particular importance for making a decision that is likely to lead the vehicle into the next customer agglomeration. Altogether, the course of the importance of the presumed customers within one of the four agglomerations seems to be caused by similar dynamics as those leading to the pile-like shape in case of the EV region. With no successive agglomeration in sight, the fifth part of the parameter curve shows a relatively low peak value of 0.1 at the beginning and a smooth decline to zero afterwards.

As illustrated in Fig. 8.10b, the solution quality of PC-1 evolves from 8.164 confirmations on average to a level of approximately 8.2 confirmations within the first 500 iterations. From iteration 500 on the solution quality shows only minor changes and does not exceed the level of 8.2 significantly. Thus, PC-1 exceeds RW by 0.4% only. Similar results appear for the remaining nine ERC-sets comprising twenty customers in the CL region. The relatively small improvements of PC-1 with respect to RW may partly be explained by the fact that according to PC-1 most of the parameters are assigned low values which are quite close to the RW parameter values of zero. For the exemplary ERC-set under investigation, 57% of the PC-1 parameter values are below 0.1 and a total of 79% of the values are below 0.2.

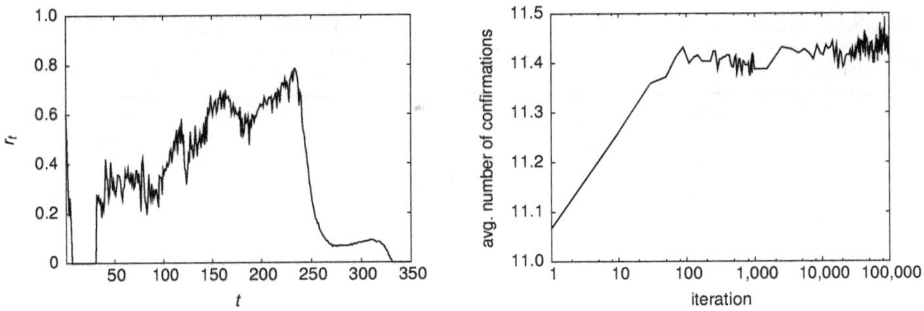

Fig. 8.11a Parameter values of PC-1 after 500,000 iterations subject to five early requests in the CL region.

Fig. 8.11b Evolution of the solution quality of PC-1 subject to five early requests in the CL region.

However, the relatively small improvements must not be exclusively attributed to the geographical distribution of customer locations in the CL region. Rather they are induced by the combination of a high number of early requests and the CL region. For comparison, Fig. 8.11 shows both parameter values and the evolution of the solution quality for an exemplary set of only five early requests in the CL region. Evidently, for by far most of the decision times, the parameter values displayed in Fig. 8.11a are significantly larger than they typically are for twenty early requests in the CL region. This suggests that the tight schedule imposed by twenty early requests in the CL region does not leave much room for improvements with respect to RW.

In contrast, the evolution of the solution quality presented in Fig. 8.11b shows considerable improvements with respect to RW in the case of five customers. For the ERC-set considered, the quality evolves from 11.07 up to nearly 11.5 confirmations on average, which corresponds to a 3.83% improvement with respect to RW and a 6.88% improvement with respect to PC. The corresponding parameter curve displayed in Fig. 8.11a shows more ruggedness and less prominent relapses than the shape of Fig. 8.10a. The sawtooth-like shape still dominates, however all things considered, the curve is likely to represent an overlay of the sawtooth-like shape and the pile-like shape observed within the previous examples.

In summary, a number of principal conclusions can be made about the properties of solutions generated by PC-1:

- Evenly distributed customer locations lead to a pile-like shape of the parameter curve, while clustered customer locations provoke a sawtooth-like shape. The experiments conducted with semi-clustered customer locations yield parameter curves that basically resemble the pile-like shape. Depending on the specific ERC-set, this shape is slightly modified by small intermediate peaks.
- On each of the service regions, the parameter values tend to go down to a lower level on average, if the number of early requests is increased, i.e., the presence of more early requests increases the focus on short-term confirmations.

Table 8.13 Average solution quality of PC-1 subject to five early requests in the EV region.

ERC set	PC-1	Δ DW in %	Δ RW in %	Δ LW in %	Δ PC in %
1	14.78	8.71	3.01	1.68	9.03
2	14.92	10.16	2.42	1.11	10.36
3	15.23	8.71	3.20	1.90	10.67
4	15.26	9.75	3.11	0.95	14.39
5	14.29	9.53	2.26	0.79	6.66
6	14.20	10.98	2.63	0.85	4.08
7	14.78	10.53	2.41	1.87	19.23
8	14.36	8.20	2.48	0.73	7.84
9	15.47	8.56	2.85	1.47	10.20
10	14.55	9.25	2.77	0.39	12.25
mean	**14.78**	**9.44**	**2.71**	**1.17**	**10.47**
STD	**0.44**	**0.92**	**0.32**	**0.53**	**4.21**

- On each of the service regions, a larger number of early requests increases the ruggedness of the parameter curve at the beginning of the time horizon. In particular, both intensity and duration of the ruggedness increase if more early requests are present.
- PC-1 exceeds the solutions qualities of both PC and RW significantly. The only exception from this rule occurs in the case of twenty early requests in the CL region. In this case PC-1 hardly exceeds the solution quality of RW.

8.3.2 Results

In addition to the representative examples considered in the previous section, the present section provides a general comparison of the results of PC-1 with the results produced by non-reactive anticipation. In particular, the solution quality of PC-1 is contrasted with the results of RW and PC as well as with the results of DW and LW. WS is omitted, because it performs worse than any other approach. Moreover, CW is not considered, because its average performance never exceeds both LW and DW.

Detailed results of PC-1 for the case of five early requests in the EV region are provided in Table 8.13. The second column of the table indicates the average number of confirmations PC-1 yields for the same ten ERC-sets as considered in Tables 8.5 and 8.12. Again, the averages are generated by applying the supposedly best parameter set within the first 100,000 iterations to the same 250 test trajectories as used for evaluation purposes in Sect. 8.2. The four rightmost columns of Table 8.13 show the relative performance of PC-1 with respect to DW, RW, LW and PC respectively.

Altogether, the experiments lead to a sample mean of 14.78 confirmations on average and a sample standard deviation of 0.44. Moreover, PC-1 outperforms each of the non-reactive approaches on each of the ERC-sets considered. The most

Table 8.14 Average solution quality of PC-1 subject to twenty early requests in the EV region.

ERC set	PC-1	Δ DW in %	Δ RW in %	Δ LW in %	Δ PC in %
1	9.44	5.07	6.21	0.25	10.12
2	8.35	3.42	7.36	0.14	13.67
3	7.32	4.81	8.28	2.63	9.05
4	7.31	4.52	11.33	-0.76	5.85
5	7.56	4.31	11.84	6.54	4.89
6	7.95	5.63	9.83	1.64	11.87
7	6.92	3.84	8.19	2.91	15.86
8	7.76	4.58	7.66	-0.05	9.98
9	7.91	3.67	8.15	-1.98	7.73
10	6.56	2.63	10.74	-1.62	12.25
mean	**7.71**	**4.25**	**8.96**	**0.97**	**9.34**
STD	**0.80**	**0.88**	**1.87**	**2.55**	**3.78**

significant gap arises between PC-1 and DW, as PC-1 exceeds DW by 9.44% on average with a sample standard deviation of only 0.92%. The average difference between PC-1 and PC even tops the gap between PC-1 and DW by more than one percent. However, the corresponding sample standard deviation is at the higher level of 4.21% and the surplus of PC-1 with respect to PC oscilliates between a minimum of 4.08% and a maximum of 19.23%. Both RW and LW are outperformed by PC-1 with very low sample standard deviations. On average, PC-1 exceeds the results of LW by 1.17% subject to a standard deviation of 0.53%. Moreover, it tops the results of RW by 2.71% on average with a standard deviation of 0.32%.

A significant improvement of the quality of PC-1 with respect to RW is provoked by an increased number of early requests. As illustrated in Table 8.14, PC-1 outperforms RW by 8.96% on average if twenty early requesting customers are present in the EV region. In addition, the table shows that twenty early requests sometimes lead to advantages of LW. While PC-1 still outperforms RW, DW and PC on each of the ERC-sets considered with twenty customers, it is superior to LW in only six out of ten cases. However, PC-1 still exceeds LW by approximately one percent on average. In the worst case, PC-1 shows a deficit of 1.98% with respect to LW, while in the best case, it outperforms LW by 6.54%. Thus, in the case of twenty early requests, PC-1 may yield larger benefits with respect to LW than it does with respect to DW. DW is exceeded by 4.25% on average and again shows a quite low sample standard deviation of the relative performance. As in the case of five early requests, both the best relative performance and the highest sample standard deviation of this performance occur with respect to PC. PC-1 outperforms PC by 9.34% on average, with single improvements ranging from 4.89% up to 15.86%.

The sample means for ten early requests in the EV region basically represent intermediates of the corresponding sample means for five and twenty early requests. Table 8.15 illustrates that merely the sample means with respect to LW do not show a clear trend. However, the performance on the single ERC-sets suggests that a higher number of early requests generally reduces the dominance of PC-1 with respect to

Table 8.15 Average solution quality of PC-1 subject to the EV region.

EV region		PC-1	Δ DW in %	Δ RW in %	Δ LW in %	Δ PC in %
5 ERC	mean	14.78	9.44	2.71	1.17	10.47
	STD	0.44	0.92	0.32	0.53	4.21
10 ERC	mean	12.22	8.07	3.96	0.57	10.14
	STD	0.38	1.01	0.69	0.64	6.72
20 ERC	mean	7.71	4.25	8.96	0.97	9.34
	STD	0.80	0.88	1.87	2.55	3.78

LW. In the experiments with ten early requests, PC-1 outperforms LW in nine out of ten cases, which is in between ten out of ten in the five early request situation and six out of ten in the twenty early request situation. If ten early requests are present in the EV region, PC-1 is 0.5% worse than LW in one case, but exceeds the result of each other approach for each ERC-set considered.

The bottom line of the experiments concerning the EV region is that PC-1 is on average superior to every other approach. Moreover, DW, RW and PC are outperformed by PC-1 on every single considered ERC-set. Only LW sometimes shows advantages with respect to PC-1, insofar as a quite large number of early requesting customers is considered.

Similar conclusions may be drawn for the SC region. As in the EV region, PC-1 outperforms DW, RW and PC on each of the ERC-sets considered subject to semi-clustered customer locations. Table 8.16 shows that on the SC region the superiority of PC-1 increases with respect to PC, and decreases with respect to both DW and RW. Nevertheless, both DW and RW are still outperfomed significantly.

In contrast, the difference between PC-1 and LW reduces to a minimum. Already in the five early request situation both approaches yield only slightly different results. PC-1 is superior on nine out of ten ERC-sets but only produces an average surplus of 0.29% compared to LW. If the number of early requests is increased, this advantage vanishes completely. For both ten and twenty early requests the solution qualities produced by PC-1 and LW are basically equal. The experiments with the ERC-sets show that none of the two approaches exceeds the other by more than 0.6% if ten early requests are present and by more than one percent if twenty early requests are present. Consequently the strong similarity of PC-1 and LW is also reflected in the quite small sample standard deviations of their relative differences.

The similarity of PC-1 and LW reduces dramatically as soon as the CL region is considered. Additional information about the case of five early requests is provided in Table 8.17. This table illustrates that PC-1 outperforms LW as well as RW and PC on each of the same ten ERC-sets as considered in Table 8.7. PC turns out to be worst, with PC-1 exceeding it by at most 15.73% and by 8.38% on average. LW performs better than PC, but is exceeded by PC-1 by at most 7.29% and by 2.86% on average. The results of RW are slightly better than the results of LW, nevertheless RW is outperformed by PC-1 by a maximum of 5.12% and by 2.80% on average.

Table 8.16 Average solution quality of PC-1 subject to the SC region.

SC region		PC-1	Δ DW in %	Δ RW in %	Δ LW in %	Δ PC in %
5 ERC	mean	15.46	3.73	1.76	0.29	14.29
	STD	0.48	0.88	0.32	0.38	4.97
10 ERC	mean	13.32	3.11	1.94	-0.06	11.99
	STD	0.58	0.94	0.52	0.38	3.28
20 ERC	mean	9.01	2.82	3.17	-0.02	13.28
	STD	0.86	0.56	0.91	0.58	4.86

Table 8.17 Average solution quality of PC-1 for five early requests in the CL region.

ERC set	PC-1	Δ DW in %	Δ RW in %	Δ LW in %	Δ PC in %
1	12.11	-5.20	2.99	1.71	7.34
2	11.64	-4.97	4.08	3.08	7.26
3	12.22	-2.46	1.53	5.53	10.09
4	11.47	0.10	2.10	2.39	4.79
5	11.18	2.19	1.97	1.01	15.73
6	11.94	2.72	1.36	0.81	8.94
7	11.49	1.99	3.83	2.64	6.88
8	11.73	0.45	1.63	1.35	6.46
9	12.13	-7.59	3.41	7.29	10.29
10	12.06	-6.97	5.12	2.79	5.97
mean	**11.80**	**-1.97**	**2.80**	**2.86**	**8.38**
STD	**0.35**	**3.96**	**1.28**	**2.06**	**3.13**

DW and PC-1 yield a draw situation with respect to the number of ERC-sets dominated. Although each of the two approaches is best on five ERC-sets, DW proves to be ahead by 1.92% on average. In the worst case PC-1 is 7.59% worse than DW whereas at the same time PC-1 is ahead of DW by a maximum of 2.72%.

Nevertheless, the superiority of DW decreases as soon as the number of early requesting customer grows. Table 8.18 contrasts the overall results of Table 8.17 with the overall results subject to ten and twenty early requests respectively. It shows that for ten early requests, DW is on average 1.2% ahead of PC-1. With this number of early requests, PC-1 again outperforms RW, LW and PC on average. PC-1 is superior to both RW and PC on each of the ten customer ERC-sets, whereas it exceeds the result of LW in eight out of ten cases. However, the average solution qualities achieved by both LW and RW are closer to the average solution quality of PC-1 than they were in the case of five early requests.

This tendency towards assimiliation is strongly reinforced, if the number of early requests is increased from ten to twenty. Table 8.18 shows that for twenty early requests the average solution quality of PC-1 is only 0.33% above the average solution quality of RW as well as merely 0.37% below the solution quality of LW. The fact that PC-1 hardly exceeds RW was already discussed and illustrated in Sect. 8.3.1

Table 8.18 Average solution quality of PC-1 subject to the CL region.

CL region		PC-1	Δ DW in %	Δ RW in %	Δ LW in %	Δ PC in %
5 ERC	mean	11.80	-1.97	2.80	2.86	8.38
	STD	0.35	3.96	1.28	2.06	3.13
10 ERC	mean	9.96	-1.20	1.96	0.85	19.95
	STD	0.60	4.26	1.69	2.90	11.01
20 ERC	mean	7.36	1.40	0.33	-0.37	18.08
	STD	0.91	1.35	0.20	2.49	5.72

(cf. Fig. 8.10b). It is a unique characteristic of the twenty early request situation in the CL region.

A similarly close alignment of the average solution qualities of PC-1 and LW can also be observed in the SC region (cf. Table 8.16). However, on the CL region the relative performance of PC-1 with respect to LW is subject to a quite high sample standard deviation. In particular, the experiments with the twenty customer ERC-sets show the performances of PC-1 and LW to differ by up to 4.3%. This clearly suggests the two approaches to implement different behaviors.

In summary, in the case of twenty early requests in the CL region PC-1 does not exceed RW significantly whereas for some ERC-sets a more successful approach (LW) than PC-1 and RW can be formulated. Thus, in the case of twenty early requests in the CL region, the contribution achieved by relying on Hypothesis 1 is only marginal although there is a potential for a greater improvement. Moreover, the fact that the average performances of PC-1, LW, RW and DW are very similar in this case suggests that a general strategy for beneficial allocation of waiting time may be very hard to achieve.

The following conclusions can be drawn concerning the performance of PC-1:

- In the EV region PC-1 outperforms every other approach on average independent of the number of early requests. It dominates DW, RW and PC on every ERC-set considered. Moreover, PC-1 exceeds LW on every five customer ERC-set. It dominates LW on nine out of ten ERC-sets with ten early requests and in six out of ten sets with twenty early requests.
- In the SC region PC-1 outperforms DW, RW and PC on every ERC-set considered and it achieves the same average performance as LW independent of the number of early requests. In addition, the sample standard deviations of the relative performances of PC-1 and LW are very low, which suggests that the behaviors implemented by PC-1 and LW are quite similar.
- In the CL region PC-1 outperforms RW and PC on every ERC-set considered. In addition, it dominates LW for both five and ten early requests and shows on average approximately the same performance as LW for twenty early requests. With respect to DW, PC-1 is on average ahead for twenty early requests, but performs only second-best in the presence of both five and ten early requests.

Altogether PC-1 shows advantages with respect to each of the remaining approaches. It is always superior to both RW and PC. DW shows to be competitive merely for a low number of early requests in the CL region. However, even in the case of five early requests it outperforms PC-1 only on half of the ERC-sets and by less than two percent on average. LW exceeds PC-1 for some ERC-sets comprising a high number of customers. However, a surplus of more than two percent can only be observed for a fraction of the twenty customer ERC-sets in the CL region.

On top of everything, there is the insight that PC-1 does not lead to significant improvements with respect to RW in the presence of twenty early requests in the CL region as well as the insight that a general improvement is quite hard to achieve in this case. A more detailed investigation of this particular case is presented in the following section.

8.4 Fine-grained Value Function Approximation

Essentially the fine-grained value function approximation represents a generalization of the elementary value function approximation. It relies on the assumption that the marginal value of a presumed customer does not only depend on the current point in time but also on the vehicle's current target location (cf. Sect. 7.2.1). Whereas this generalization promises additional benefits in any case, it is most relevant with respect to the most challenging problem instances.

The experiments of the previous section suggest that dynamic routing of a service vehicle is particularly challenging in the presence of a relatively large number of early requests in a service region with customer agglomerations. Remember the case of twenty early requests in the CL region. For this scenario RW, DW, LW and PC-1 yield similar results on average although they pursue different principles. The corresponding parameter values of PC-1 indicate that considering the presumed customers is of particular importance as soon as the vehicle tends to jump into the next customer agglomeration. However, at the same time, PC-1 does not lead to a significant improvement of the quality of RW. The cause of PC-1 not being able to convert its parameter values into more confirmations may be the lacking precision of the elementary value function approximation. A more efficient routing strategy may require a more fine-grained approximation for exploiting the small tolerance of the tight schedule that typically occurs with twenty early requests in the CL region.

Thus, the focus of this section is on approximate anticipation by means of the fine-grained value function approximation. In analogy to the denotation of PC-1, the corresponding approach is referred to as "PC-2" in the following. Section 8.4.1 analyzes the behavior of PC-2 for the case of twenty early requets in the CL region. Subsequently, Sect. 8.4.1 provides additional insights about PC-2 by considering variations of the actor-critic configuration imposed in Sect. 8.1.2.

Table 8.19 Average solution quality of PC-2 for twenty early requests in the CL region.

ERC set	PC-2	Δ DW in %	Δ RW in %	Δ LW in %	Δ PC in %
1	8.55	0.56	1.42	-1.25	8.70
2	7.00	5.68	4.17	6.26	19.14
3	6.36	5.37	2.45	4.13	14.55
4	7.74	5.80	5.28	0.47	26.99
5	8.72	1.77	1.63	-0.46	27.65
6	8.10	4.87	2.63	3.84	18.90
7	6.42	6.15	3.81	-0.37	21.30
8	6.62	0.61	1.16	-0.24	19.67
9	8.38	3.25	2.65	4.38	27.98
10	7.40	4.46	2.27	3.52	23.81
mean	**7.53**	**3.85**	**2.75**	**2.03**	**20.87**
STD	**0.90**	**2.16**	**1.31**	**2.66**	**6.14**

8.4.1 Results and Solution Properties

Table 8.19 illustrates the performance of PC-2 for the same ten sets of twenty early requests in the CL region as considered in the previous sections. First of all, the table shows that PC-2 clearly outperforms every non-reactive approach on average. PC-1 is exceeded as well, because it performs very similar to RW for each of the twenty customer ERC-sets in the CL region (cf. Table 8.18 and Fig. 8.10b in Sect. 8.3.1).[5]

Just as PC-1, the three non-reactive approaches DW, RW and PC are outperformed by PC-2 on every single ERC-set considered. In contrast, PC-2 is superior to LW in only six out of ten cases. However, a closer look at the results shows that LW is ahead by more than one percent only once, whereas PC-2 is superior to LW by more than three percent in five cases. In the best case PC-2 even exceeds LW by more than six percent contributing to an average surplus of slightly more than two percent.

An average surplus of 3.85% occurs with respect to DW, although the best performance of PC-2 with respect to DW even implies a marginally lower surplus than the best performance with respect to LW. Expectedly, PC-2 always is strongly superior to PC. Nevertheless, the fact that the former exceeds the latter by more than twenty percent on average is of interest, because it illustrates the potential loss of profits due to an inadequate assumption on the marginal value of a presumed customer.

Above all, the superiority of PC-2 with respect to RW shows that taking into account the presumed customers leads to significant benefits also in the case of twenty early requests in the CL region.

As an example, Fig. 8.12 displays the evolution of the solution quality of PC-2 for the same ERC-set provoking the evolution of the solution quality of PC-1 shown

[5] Due to the strong similarity of the results of RW and PC-1, only the results of the former are included in Table 8.19.

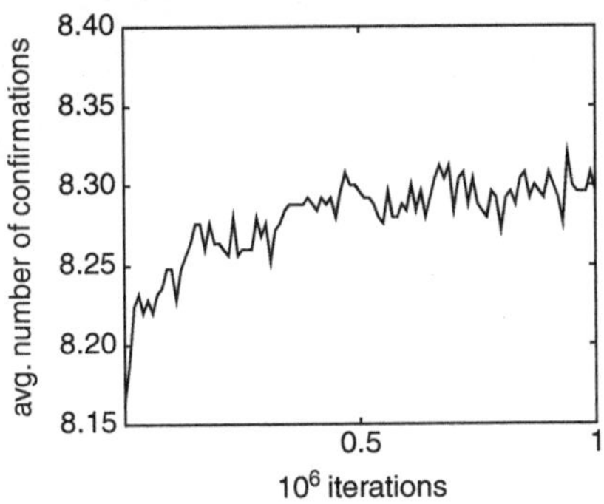

Fig. 8.12a Evolution of PC-2 from iteration one to iteration one million subject to twenty early requests in the CL region.

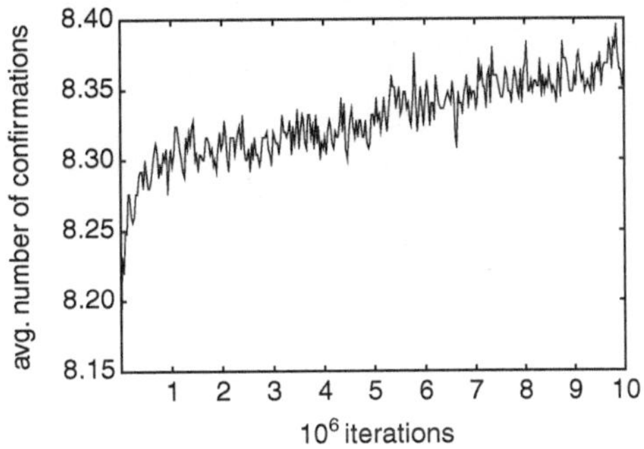

Fig. 8.12b Evolution of PC-2 from iteration one to iteration ten million subject to twenty early requests in the CL region.

in Fig. 8.10b. Both PC-1 and PC-2 start from a level of approximately 8.16 confirmations on average. However, PC-1 hardly exceeds 8.20 confirmations, whereas the solution quality of PC-2 climbs up to nearly 8.40 confirmations within ten million iterations. Figure 8.12a illustrates that PC-2 hits the level of 8.20 confirmations already after 15,000 iterations and then evolves up to a level of 8.30 within the first million iterations. Beyond iteration one million, the solution quality initially seems to oscillate at a quite stable level. As shown in Fig. 8.12b, the start of the next phase of significant improvement may then be identified at three million iterations, from which on the solution quality trends to rise up to a level of nearly 8.40 at ten million iterations. Although the evolution suggests that further improvements will occur in the upcoming iterations, the total number of iterations for generating the results of Table 8.19 is limited to ten millions in order to keep the computational burden at a reasonable level.[6]

The computational burden is not so much an issue with PC-1, which however cannot convert its parameter values into a significant improvement of the solution quality for twenty early requests in the CL region. Primarily, the parameter curves generated by PC-1 provide valuable insights, as illustrated by the discussion of the sawtooth-like parameter curve shown in Fig. 8.10a of Sect. 8.3.2. According to this discussion, for example $t = 140$ must be considered as a point in time at which the vehicle is about to make a transition between two customer agglomerations. In particular, the fact that the parameter value $r_{140} \approx 0.28$ of PC-1 is close to the second peak of the sawtooth-like shape (cf. Fig. 8.10a) indicates that the current transition leads from the second towards the third customer agglomeration visited.

As opposed to the parameters of PC-1, the parameters of PC-2 additionally provide information about the order in which the customer agglomerations are visited by the vehicle. Due to the tight schedule imposed by twenty early requests in the CL region, PC-2 updates for each point in time only the values of those parameters which are associated with a geographical location that is relatively close to the

---

[6] See Sect. 8.4.2 for details on this issue.

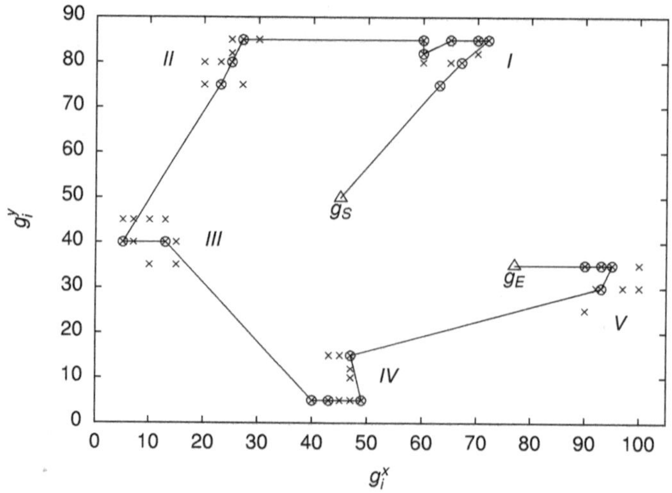

Fig. 8.13 Planned route at $t = 0$.

current position of the vehicle. Thus, the subset of locations receiving at least one update for a certain point in time t indicates the part of the service region in which the vehicle typically remains at t.

A closer look at the parameter values after ten million iterations unveils that for the given ERC-set, the vehicle typically visits the customer agglomerations in the same order as induced by the planned route at $t = 0$. As displayed in Fig. 8.13, this route traverses each of the five agglomerations exactly once. Initially, the vehicle moves to the upper right corner of the service region and typically visits an early requesting customer[7] at first. After serving part of the remaining customers in agglomeration *I*, the vehicle moves on to the upper left corner of the service region for visiting the customers of agglomeration *II*. Subsequently, transitions occur towards agglomerations *III*, *IV* and *V* before the vehicle finally goes for the end depot.

Figure 8.14 indicates that, just as in the case of PC-1, the transition from agglomeration *II* to agglomeration *III* occurs around $t = 140$. The figure displays the CL region together with the parameter values of PC-2 at $t = 140$. Customer locations, whose corresponding parameters r_{140}^n did not receive a single update within ten million iterations are represented as crosses, whereas the customer locations that did receive updates are represented as circles. The latter may be considered as the set of locations the vehicle is most likely to move to (or wait at) at $t = 140$. Of course, the fact that this set comprises locations within both agglomeration *II* and agglomeration *III* does not prove that a transition may occur at $t = 140$. However, this distribution of locations may be considered as a stong indication that the transition from *II* to *III* typically occurs around $t = 140$.

As shown in Fig. 8.14 a total of sixteen parameters with $t = 140$ received an update up to iteration ten million. The values of these parameters range from slightly

[7] The twenty early requesting customers are highlighted by circles within Fig. 8.13.

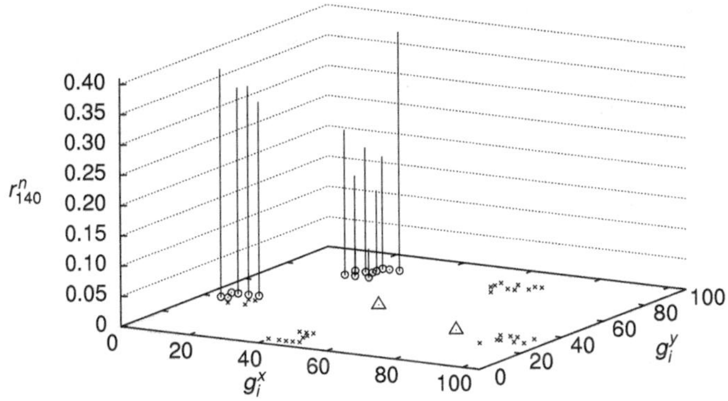

Fig. 8.14 Parameter values of PC-2 at $t = 140$.

more than zero to approximately 0.4. At a more detailed level, three categories of values may be distinguished. First of all, most of the affected locations within agglomeration *III* as well as the rightmost location of agglomeration *II* are associated with relatively high parameter values ranging from 0.32 to 0.39. Moreover, five of the locations within agglomeration *II* are associated with moderate parameter values between 0.13 and 0.23. Each of the parameters within these two categories of values received several thousand updates. On the contrary, the third category comprises six parameters with values close to zero that are generated by only a few hundred updates each.

Closer inspection of the evolutions of the parameter values indicates that almost none of the sixteen values at $t = 140$ has converged up to iteration ten million. However, the evolutions suggest that beyond this iteration major changes will primarily occur with respect to the parameters currently being at a value close to zero. Most likely, the two categories of high and moderate values will persist.

Recall that PC-1 proposes a single parameter value of 0.28 at $t = 140$. Against this background, the fact that Fig. 8.14 shows a broad range of significantly different values at $t = 140$ provides an explanation of the improved performance of PC-2 with respect to PC-1. Evidently, the elementary value function approximation of PC-1 does not allow for enough degrees of freedom in order to capture the different post-decision states at $t = 140$ effectively.

Due to the various influences involved, a detailed explanation of the single parameter values of PC-2 is hardly possible. Nevertheless a number of observations can be made. Within agglomeration *II* the particularly large value associated with the rightmost customer location catches one's eye. It shows that the value of a decision to either move to this location or wait there, strongly depends on the presumed customers. On the contrary, the rather moderate parameter values associated with the remaining locations of the agglomeration indicate that the presumed customers generate only a minor contribution to the value of a decision to move or wait there. In general, a parameter value within agglomeration *II* tends to be lower, if it is

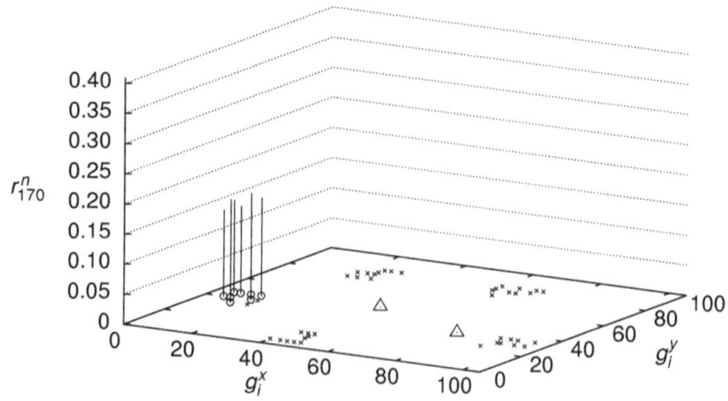

Fig. 8.15 Parameter values of PC-2 at $t = 170$.

associated with a location in the center of the agglomeration instead of with a location in an area close to a neighboring agglomeration.

In particular, the category of relatively high parameter values seems to be induced by the pending transition to the next agglomeration. This assumption is supported by the dominance of high parameter values within agglomeration *III* as well as by the situation illustrated in Fig. 8.15. The figure shows the parameter values of PC-2 at $t = 170$, which is a point in time at which the vehicle is very likely to remain within agglomeration *III*. A transition to a successor agglomeration will probably not occur, because each of the eight candidate locations for a vehicle move at $t =$ 170 are within agglomeration III. At the same time, the values of the parameters at $t = 170$ are both moderate and quite similar to each other. Besides a single parameter value of 0.07, each of the values at $t = 170$ is between 0.14 and 0.18. A category of relatively high parameter values as occuring at $t = 140$ cannot be observed.

The different parameter profiles of Figs. 8.14 and 8.15 turn out to be examples of two oppositional situations typically occurring in the case of twenty early requests in the CL region. Either there is no pending transition as in the example of $t = 170$ or there is a transition to come as in the example of $t = 140$. In the former case, the parameter values are both moderate and similar to each other, whereas in the latter case different categories of parameter values occur. As a consequence, the single parameter value derived by PC-1 is of low effectiveness in case of a pending transition, whereas such a value tends to be sufficient if the vehicle remains within its current agglomeration. Note that in the case of $t = 170$, PC-1 proposes a value of 0.16, which implies a relatively low average deviation from the correpsonding values of PC-2. However, the fact that these values are relatively low, reduces their impact on the improvement of the overall performance with respect to RW.[8] At the same time this underlines that the major impact on the improvement of the overall performance is exerted by decisions made subject to a pending transition, which are supported adequately by the parameter values of PC-2.

[8] Recall that RW assumes every parameter value to be zero.

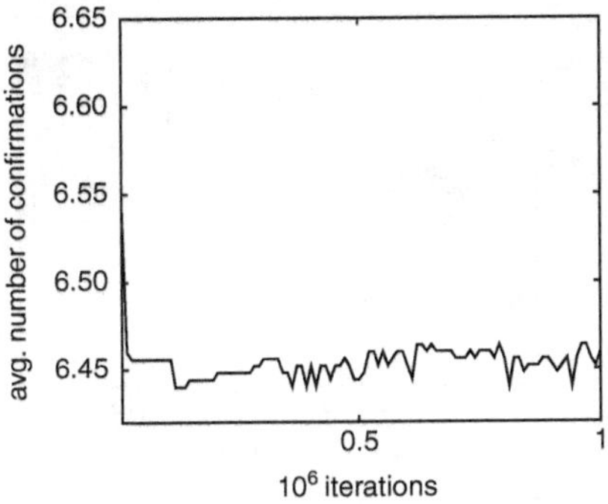

Fig. 8.16a Evolution of the solution quality of PC-2 showing an initial drop-off. (Iteration one to iteration one million subject to twenty early requests in the CL region.)

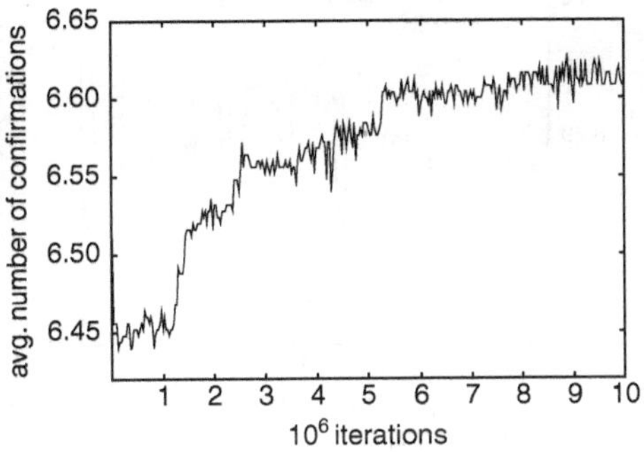

Fig. 8.16b Evolution of the solution quality of PC-2 showing an initial drop-off. (Iteration one to iteration ten million subject to twenty early requests in the CL region.)

In summary, the fine-grained value function approximation of PC-2 enables a successful routing strategy for twenty early requests in the CL region. Whereas the competing approaches perform similar on average, PC-2 exceeds their solution qualities, because it is capable of exploiting the small tolerance of the tight schedule induced by twenty early requests in the CL region. In particular, PC-2 successfully handles the situations with a pending transition to the next agglomeration of customers. Nevertheless, compared to PC-1, a quite large number of iterations of the actor-critic method may be required in order to realize the full potential of PC-2. An extreme case is displayed in Fig. 8.16, which shows the evolution of the solution quality of PC-2 for another set of twenty early requests in the CL region. Two specific features of the evolution are exposed in Fig. 8.16a. First of all, a significant drop-off of the solution quality can be observed within the initial iterations. Starting from a number of 6.54 confirmations on average, the solution quality drops down below 6.45 confirmations within the first hundred thousand iterations. Thereafter the solution quality hardly recovers until iteration one million. Figure 8.16b illustrates that significant improvements do not occur until approximately iteration 1.2 million, from which on the solution quality trends to increase continuously.

The example of Fig. 8.16 indicates that the large number of parameters considered for the fine-grained value function approximation introduces a higher level of sensitivity of the initial policy, i.e., the initial policy is less robust with respect to changes of the parameter values in the early iterations. Evidently, this higher level of sensitivity strengthens the need for a larger number of iterations of the actor-critic method in order to realize the full potential of PC-2. As illustrated within the following section, an influence on the required number of iterations may be exerted by means of variation of the actor-critic configuration identified in Sect. 8.1.2.

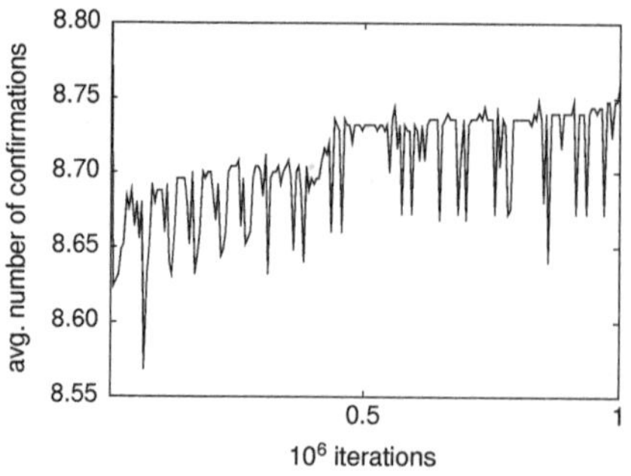

Fig. 8.17a Evolution of the solution quality of PC-2 from iteration one to iteration one million subject to twenty early requests in the EV region.)

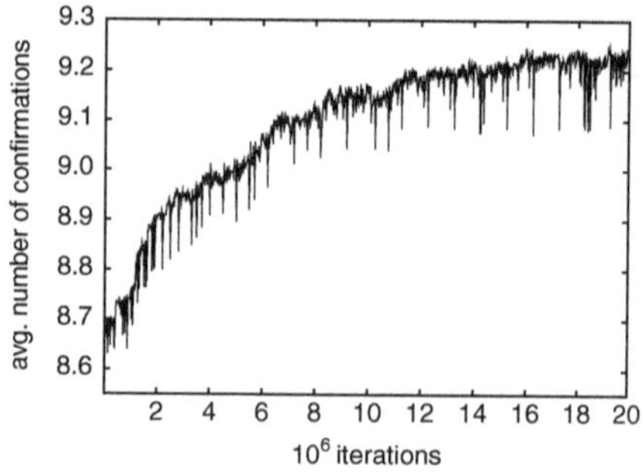

Fig. 8.17b Evolution of the solution quality of PC-2 from iteration one to iteration twenty million subject to twenty early requests in the EV region.)

8.4.2 Variations

The preceding section demonstrates that realizing the full potential of the fine-grained value function approximation requires an increased number of actor-critic iterations. Of course, the number of iterations is not so much an issue if there are no rigorous time constraints on the learning phase. For example, if a company with a fixed number of registered customers encounters exactly the same initial situation at the beginning of every shift, a learning phase comprising even hundreds of millions of iterations may be executed once and the resulting policy may be applied repeatedly. However, in many cases, either the set of early requests or the customers' request probabilities are subject to fluctuation and the amount of computing time left until the beginning of the next shift is limited. In such a case, both increased computational efficiency and the availability of computing power may be of particular relevance.

First of all, a larger number of iterations per time unit may be achieved by computer hardware upgrades and by advanced software engineering techniques such as multi-threading and parallelization. Without such measures and without code optimization, one million iterations of the actor-critic method applied in the preceding section require approximately twenty hours of runtime on a desktop PC with a 2.8GHZ CPU. Basically, the previously mentioned technical measures should be sufficient for reducing the runtime down to a level permitting for example overnight calculations.[9] Nevertheless, the technical effort may be contained by increasing the efficiency of the actor-critic framework. In order to give an impression of the possible starting points for such an increase, two variations of the the actor-critic configuration defined in Sect. 8.1.2 are outlined in the following.

[9] Think of the potential benefits of parallelization, especially if more than one trajectory is considered for policy evaluation.

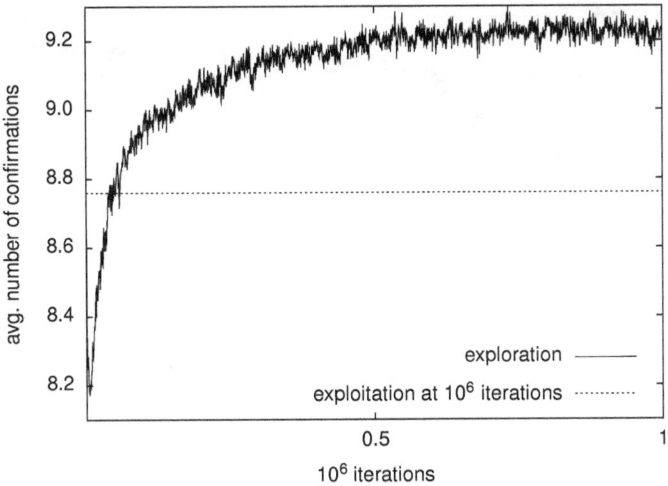

Fig. 8.18 Example evolution of the solution quality of PC-2 subject to twenty early requests in the EV region using state space exploration.

Exploiting the possibilities of the actor-critic framework is of particular importance in the face of an evolution of the solution quality as shown in Fig. 8.16b for the case of twenty early requests in the CL region. Whereas both an initial drop-off of the solution quality and a phase of slow recovery do not always occur, there also are problem settings for which those two features are particularly strong. Such an example is provided in Fig. 8.17. The figure displays the evolution of the solution quality of PC-2 for a set of twenty early requests in the EV region. As shown in Fig. 8.17a, within the very first iterations the solution quality drops down from an average of 8.8 confirmations to roughly 8.63 confirmations on average. The solution quality then recovers quite slowly and finally reaches a level of slightly more than 8.75 at one million iterations. Figure 8.17b illustrates that the initial level of 8.8 confirmations is reached again at approximately 1.5 million iterations and that the solution quality trends to improve continuously up to a level of 9.24 at twenty million iterations.

Figure 8.18 shows that approximately the same level can be achieved already after one million iterations. To this end, the actor-critic configuration identified in Sect. 8.1.2 must be modified with respect to the state sampling procedure. In particular, Fig. 8.18 displays the evolution of the solution quality of PC-2 relying on exploration state sampling (cf. Sect. 7.1.1). Evidently, state space exploration has a strong impact on both initial drop-off and recovery. On the one hand, the solution quality drops down from a level of 8.8 to less than 8.2 confirmations on average within the initial iterations. On the other hand, however, the solution quality recovers much more quickly than with exploitation state sampling and hits the level of 8.8 again after approximately 50,000 iterations. Beyond this iteration, the solution quality improves quite smoothly and finally remains at a level of 9.23 confirmations at one million iterations, i.e., for the problem setting considered, changing the

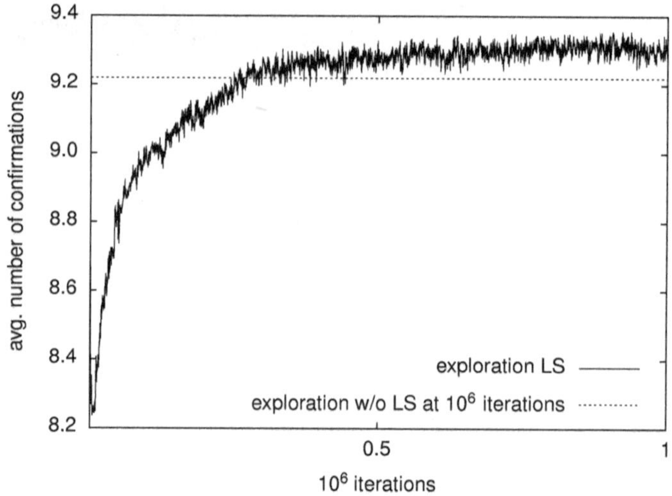

Fig. 8.19 Example evolution of the solution quality of PC-2 subject to twenty early requests in the EV region using state space exploration and local search.

state sampling procedure reduces the number of actor-critic iterations for getting to a level of slightly more than 9.2 by 95%.

Nevertheless, state space exploration also shows certain drawbacks. As illustrated in Sect. 7.1.1, an actor-critic iteration with state space exploration entails more computational effort than an actor-critic iteration with exploitation state sampling. Exploration increases the total effort, because it requires execution of an additional partial trajectory for each of the states visited within an iteration. At the same time, however, execution of the partial trajectories of an iteration suggests itself for parallelization, such that state space exploration does not necessarily lead to an increased runtime. Anyhow, the drawbacks of state space exploration raise the need for more efficient sampling strategies. An overview of elementary approaches towards more efficiency is provided by, e.g., Powell (2007, Chap. 10).

In addition to different state sampling strategies, the family of actor-critic methods defined in Sect. 7.2.4 allows for variations with respect to decision model identification. Figure 8.19 exemplifies the effect such a variation may exert on the evolution of the solution quality. It shows the evolution of the solution quality of a variant of PC-2 for the same ERC-set as considered in Fig. 8.18. The variant of PC-2 not only implements exploration state sampling, but also consults more features of Procedure 11 for verification of the feasibility of a candidate decision. Actually, verification of feasibility still relies on a single GRASP-iteration only, but additionally consults the local search operator. As a consequence, the ratio of identified feasible decisions to actually feasible decisions is likely to increase. Moreover, as shorter routes lead to larger amounts of slack, the more elaborate approach to determination of route lengths exerts an influence on the value function approximation.

Figure 8.19 shows that the major result of these changes is a further reduction of the actor-critic iterations needed for achieving a level of 9.23 confirmations on average. Already after approximately 250,000 iterations, PC-2 with exploration and local search catches up with the performance PC-2 with exploration and without local search showed at iteration one million. Note that the former yields both a reduced drop-off and faster recovery with respect to the latter. In addition, conducting local search enables a significantly better solution quality after one million iterations. In contrast to the level of 9.23 confirmations reached by exploration without local search, using local search leads to a level of 9.31 confirmations at iteration one million.

However, just as in the case of state space exploration, the reduced number of actor-critic iterations resulting from a more elaborate version of Procedure 11 must be contrasted with an increased computational effort required for a single iteration. As a decision cannot be made without executing Procedure 11 at least once, the configuration of this procedure exerts a considerable influence on the computing time required for one iteration. A more detailed analysis of the impact of various configurations of Procedure 11 in the context of a similar actor-critic method is provided by Meisel et al (2010).

Altogether the present section illustrates that both the state sampling procedure consulted as well as the way of decision model identification heavily influence the behavior of an actor-critic method. Both issues involve opportunities for gaining additional efficiency as well as non-trivial scientific challenges.

Chapter 9
Managerial Impact of Anticipatory Optimization

The previous chapters illustrate the broad scope of anticipatory optimization. As indicated in Chaps. 1 and 2, anticipatory optimization is relevant with respect to a large variety of operational contexts. Moreover, Chaps. 3 to 5 give an impression of the tremendous methodological spectrum included in the concept. Eventually, Chaps. 6 to 8 exemplify the realization of anticipatory optimization in the context of dynamic vehicle routing. However, this example does not provide a general cookbook for successful application in arbitrary operational contexts.

Rather, the managerial impact of anticipatory optimization will be different within each context. Two general determinants may be distinguished. First of all, the impact is subject to technological preconditions. That is, both the degree of anticipation that may be realized and its resulting effect strongly depend on the technological competences and the technological infrastructure present in a company. While competences and infrastructure provide a measure for what can be realized, there also is the question of what should be realized with respect to the particular operational context under investigation. Evidently, from a managerial point of view, the selection of a degree of anticipation must meet more issues than technological feasibility only.

The following two sections provide a discussion of the general determinants of the impact of anticipatory optimization. Section 9.1 focuses on technological preconditions, whereas Sect. 9.2 covers a wider scope of issues to be considered for selecting an adequate degree of anticipation.

9.1 Technological Preconditions

Hardly an approach to anticipatory optimization may be developed successfully without the availability of probabilistic information characterizing the random influences present in the respective operational context. The need for such information is quite obvious if the actor-critic framework is involved. Moreover it is mandatory by definition for establishing probabilistic approaches to non-reactive anticipation.

Merely conventional approaches to non-reactive anticipation may basically be developed without probabilities. However, even if probabilistic information is not necessary for the design of such approaches, it is for sure desirable for the purpose of their evaluation. Without a testbed in terms of a computer simulation that implements the random influence of the environment, the performance of an approach can only be determined by means of experimental application in real business. Yet, evaluation by experimental application may imply a loss of profits. Therefore the generation of probabilistic information representing the random influence on a company's operations is recommended in any case.

Consequently, enabling technologies allowing for generation of probabilistic information are required. In the service vehicle example of the previous chapters, a company could hardly determine customer specific request probabilities without relying on information technology. The data of each customer request should be fed into a data base management system such that it can be used for further processing. Probabilities may then be derived from data aggregates which can be turned into relative frequencies. As the precision of the probabilities represents a major contribution to the overall success of an approach, a company may rather derive elaborate conditional probabilities than rely on one relative frequency per customer only. For example specific request probabilities may be derived for each day of the week or for each of the seasons of the year.

Moreover, the generation of conditional probabilities is not limited to temporal attributes only. An even higher degree of precision may be achieved by considering additional attributes characterizing either customer, company or the environment. Think of deriving probabilities subject to a customer's geographical location in combination with information about the company's regional promotion activities for example.

Note that the same careful consideration as recommended with respect to the random influences may be adequate with respect to other system attributes. For instance, travel time estimates may be consulted for deriving the values of the "distance" attributes of the road links for dynamic routing of a service vehicle. Yet, such estimates are quite likely to depend on various attributes like time of day or type of the vehicle in the field.

In many cases anticipatory optimization leads to good results even if carried out subject to a quite rough characterization of a company's operations. However, the chance of success is of course greater if an approach can rely on a detailed characterization derived from all the attributes of the operations under consideration. Consequently, the ideal support for anticipatory optimization consists in company wide data management as enabled by enterprise resource planning (ERP) systems. By means of an ERP system the data occuring in the course of a company's operations can be captured, consolidated as well as held ready for further processing. At regular intervals the data may then be transferred to an analytical database system, which provides a dedicated environment for aggregation and analysis leading to insights such as request probabilities for instance. Deriving such insights requires specific technological competences, which are typically denoted as business intelligence in the literature (e.g., Loshin, 2003).

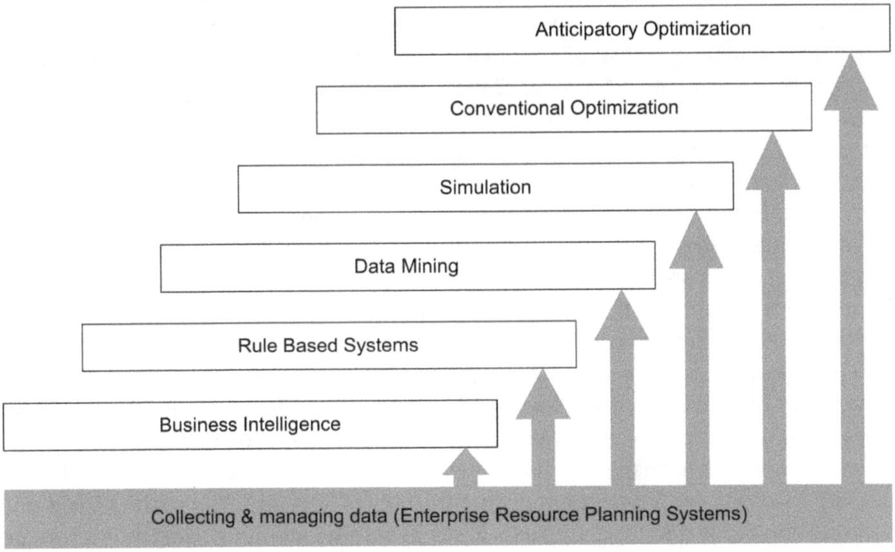

Fig. 9.1 Technological competences effecting different levels of abstraction from operational data.

Business intelligence techniques generate aggregates representing an abstraction from operational data. In the same sense, an anticipatory decision may be considered as an abstraction from operational data. However, additional technological competences are required in order to generate such an abstraction. A general classification of technological competences according to their level of abstraction from raw data is provided by Fig. 9.1.[1] The figure illustrates all the way from deriving aggregates in the sense of business intelligence to deriving decisions by anticipatory optimization. In between levels of abstraction from operational data comprise representing this data as a set of if-then rules as well as condensing the data into information structures by data mining (cf. Sect. 4.1.2). Higher levels may be achieved by transforming the raw data into a simulation environment and by deriving optimal decisions from the data as done by conventional optmization approaches (cf. Sect. 4.1.3).

Each of the levels shown hides its own complexities and each of the levels bears a distinct potential for contributing to the success of a company. However, in many cases successful realization of the potential of a higher level requires the competences of lower levels as preconditions. For example, gaining information by data mining is usually hardly possible without preceding insights achieved by business intelligence techniques such as online analytical processing. Moreover, both such insights and information derived by data mining are often necessary preconditions for establishing a decision model in the context of conventional optimization (cf. Sect. 4.3).

[1] Figure 9.1 relies on a similar figure proposed by Frans de Rooij at the 84th meeting of the GOR working group "Real World Mathematical Optimization", that took place in Duiven, The Netherlands in April 2010.

However, data mining is not always required for conventional optimization. In contrast, it is an integral part of anticipatory optimization as far as value function approximation is involved. Moreover, if the actor-critic framework is consulted, anticipatory optimization cannot be realized without a computer simulation environment for generation of state trajectories. On top of everything, conventional optimization techniques are required even if anticipatory optimization is conducted without the actor-critic framework and without value function approximation.

In summary, the full scope of anticipatory optimization may only be consulted if a number of technological competences are available. Realizing higher degrees of anticipation requires the smooth interaction of competences in the fields of business intelligence, data mining, simulation and conventional optimization. The need for those technological competences is accompanied by the need for technology in terms of infrastructure. Besides IT-systems required for data management, additional infrastructure may be necessary for both data collection and implementation of anticipatory decisions. In the service vehicle example, the former may result into the use of positioning devices for identification of the vehicle's current location, whereas the latter may lead to each driver having his communication unit for receiving updated instructions.

Thus, careful analysis of a company's technological capabilities and of the required investments is the first step towards selection of a degree of anticipation.

9.2 Selecting a Degree of Anticipation

Evidently, there are not many choices with respect to the degree of anticipation if the technological preconditions for application of the actor-critic framework are not fulfilled. However, even if each of the possible degrees can be consulted, making the right choice is not always trivial.

First of all, going for the highest possible degree of anticipation promises advantages. Maybe the most striking argument in favor of the actor-critic framework is, that it essentially effects an adaptation of decision models to the operations under consideration. As this adaptation is triggered by simulation of the operations themselves, an actor-critic method may be considered as making the operations propose how they must be controlled subject to a given set of hypotheses. Consequently an approximate value function derived by an actor-critic method is likely to represent both, a good policy and additional valuable information about the operations under consideration.

If the focus is on performance only, the benefits gained from an actor-critic approach will primarily depend on two factors. On the one hand, gaining benefits requires suitable hypotheses for value function approximation and decision model approximation respectively. The more experienced a company is with its operations, the easier both types of hypotheses can be derived. Usually, formulating a powerful and computationally efficient hypothesis for value function approximation is less straightforward than identification of a subset of important decisions. However, decision model approximation may be subject to careful consideration as well.

As an example think of the tradeoff between waiting and driving introduced within the previous chapters on dynamic routing of a service vehicle. This decision model approximation is straightforward from a mathematical point of view, but requires taking into account additional aspects. From a mathematical point of view the nature of a suboptimal move operation does not matter, i.e., with respect to the overall performance it makes no difference whether one additional future confirmation cannot be realized due to a superfluous waiting operation or if it cannot be realized due to an early move to another location. However, it might make a difference from a managerial point of view. The latter for sure declares a superfluous waiting operation an explicit misallocation of resources. Yet, an early move operation does not necessarily fall into the same category. As such an operation obviously serves a confirmed customer it may be legitimately recognized as realizing an indirect contribution to the overall goal of the company. Consequently, an explicit focus on the tradeoff between waiting and driving calls for a preceding managerial commitment to waiting as an option.

Apart from the availability of suitable hypotheses, the second factor determining the additional benefits of an actor-critic method is given in terms of both availability and quality of non-reactive alternatives. If such an alternative relies on profound human experience and has been both engineered and applied over many years, it will often be quite challenging to develop an actor-critic method that immediately outperforms its non-reactive counterpart.

Moreover, a significant performance increase by means of either implicit or approximate anticipation can of course not be expected if the non-reactive alternative in use is close to optimal. From this it follows that going for implicit or approximate anticipation is recommended if

- the technological preconditions are fulfilled,
- suitable hypothesis for approximation are available, and
- the adequacy of non-reactive alternatives is either poor or cannot be determined.

As soon as these three conditions are satisfied, the subsequent choice between implicit and approximate anticipation depends to a large extent on the nature of the hypotheses at hand. Finally, the computational resources available will be a major influence on the tradeoff between an elementary and a rather fine-grained value function approximation.

Chapter 10
Conclusions

The preceding chapters cover anticipatory optimization for dynamic decision making. Chapter 1 illustrates the characteristics of the emerging decision problems that raise the need for anticipatory optimization. It exemplifies the influences of globalization, of the progress of information technology and of the increasing resource prices on the operations of many companys. Due to these influences a large number of operations must nowadays be conducted under increasingly dynamic and uncertain conditions. Chapter 1 shows by means of examples from different contexts that such operations may be considered as dynamic decision processes.

A generic representation of dynamic decision processes is provided in Chap. 2. In particular this chapter introduces the theory of Markov decision processes and identifies the anticipatory role of optimization in this context. For the purpose of anticipation, optimization must generate decisions that do not depend on past and present only but also take into account expectations about the future. Starting from the definition of a perfect anticipatory decision, Chap. 2 introduces prototypical degrees of anticipation that may be realized by optimization. Major distinctions are made between the four degrees of "perfect", "approximate", "implicit" and "non-reactive" anticipation. Each of these degrees implies a characteristic type of optimization problem to be solved for making an anticipatory decision.

Chapter 3 exclusively focuses on perfect anticipation. It develops three consecutive approaches to deriving perfect anticipatory decisions, each of which pursues the principle of dynamic programming. The first approach is given in terms of the elementary methods of dynamic programming. These provide the foundation of the second approach, which additionally consults simulation techniques and is known as forward dynamic programming. Forward dynamic programming implies a quite large family of methods that Chap. 3 summarizes under the actor-critic framework. This framework serves as a basis of model free dynamic programming representing the third of the approaches to perfect anticipation. Concerning model free dynamic programming, the use of post-decision states proves to be of particular relevance, because this type of state may lead to a significant increase of the effectiveness of perfect anticipation.

Nevertheless, even if post-decision states are consulted, the actor-critic framework is still likely to suffer from restricted effectiveness due to both state space dimensionality and decision space complexity. These restrictions may be relaxed by introducing approximations of perfect anticipatory decision models.

Chapter 4 identifies "decision model approximation" as one of the synergies of optimization and data mining. Exploiting this synergy for effective anticipatory optimization requires data mining to generate an approximation of a perfect anticipatory objective function. As optimization and data mining basically are two paradigms independent of each other, Chap. 4 establishes a consistent conceptual foundation for illustration of their synergies. A process model is specified for optimization and data mining respectively. These models are then unified into an integral methodology unveiling the synergies of its two components. Some of the synergies serve the purpose of more efficient data mining, whereas others enable effective optimization. Effective anticipatory optimization by "decision model approximation" requires optimization and data mining to be synchronized with the actor-critic framework.

Chapter 5 elaborates on ways of synchronization. It identifies "state space aggregation" and "predictive modeling" as the primary approaches to approximate anticipation. These approaches are formulated in terms of data mining operations before they are integrated into the actor-critic framework. The resulting generalization of the actor-critic framework enables anticipatory decisions at the expense of a reduction of the degree of anticipation from perfect to approximate. Eventually, Chap. 5 points out the challenges to be faced if an actor-critic approach must be realized with respect to a particular operational context.

Such an operational context is introduced by the discussion of dynamic vehicle routing provided in Chap. 6. As dynamic vehicle routing is becoming increasingly important from an economic point of view, the field provokes a rapidly growing amount of research activities. It is established as a distinct branch of research within the general field of vehicle routing since approximately one decade.

Starting from a discussion of scientific foundations and important problem characteristics, Chap. 6 provides an overview of the state of the art of dynamic vehicle routing. In particular, contributions from the literature are classified according to their degree of anticipation. This classification unveils the fact that most of the works up to now consider only low degrees of anticipation. Although merely perfect anticipation cannot be applied to dynamic vehicle routing problems of practical relevance, approximate anticipation is hardly considered in the literature as well.

Most striking, the classification of the literature shows that there is no previous approach to approximate anticipation for dynamic vehicle routing with late requesting customers. Both the practical relevance and the tremendous complexity of this type of problem emerge from the case of dynamic routing of a service vehicle discussed at the end of Chap. 6.

Chapter 7 addresses this case by proposing a number of new anticipatory approaches to dynamic routing of a service vehicle. First of all, the actor-critic framework of Chap. 3 is considered for realizing perfect anticipation with respect to a small illustrative problem instance. Analysis of the corresponding experimental results shows that the state sampling procedure within an actor-critic method is of particular importance for getting to the optimal solution.

Due to the limited effectiveness of perfect anticipation with respect to larger problem instances, Chap. 7 consults the generalized actor-critic framework of Chap. 5 for developing an approach to approximate anticipation. Definition of adequate data mining operations as well as suitable optimization procedures eventually leads to a new family of actor-critic methods for dynamic routing of a service vehicle. Furthermore, Chap. 7 suggests a variety of approaches to non-reactive anticipation for this problem. Depending on their use of probabilistic information, these approaches are divided into the two categories of "probabilistic" non-reactive anticipation as well as "conventional" non-reactive anticipation. Part of the non-reactive procedures is established on the basis of ideas present in the literature, whereas two of them are entirely new.

Chapter 8 provides insights into the behaviors of the different procedures introduced in Chap. 7. To this end, computational experiments are conducted with respect to a number of problem instances varying in both the geographical distribution of customer locations and the number of early requesting customers. The experiments show that the relative performances of the conventional approaches heavily depend on the attributes of the problem instance at hand. Moreover, they show that on most of the instances, the conventional approaches are outperformed by competing probabilistic non-reactive approaches.

Subsequently, Chap. 8 selects a member of the family of actor-critic methods defined in Chap. 7 and shows that it applies to each of the problem instances introduced previously. Both the geographical distribution of customer locations and the number of early requests of an instance are reflected in the shape of the parameter curves generated by the actor-critic approach. Moreover, this approach shows advantages with respect to each of the competing non-reactive procedures.

The experiments suggest that dynamic routing of a service vehicle is particularly challenging in the presence of a high number of early requests and clustered customer locations. Thus, the configuration of the actor-critic method is modified such that it generates a more fine-grained approximation of value functions. As a consequence, a significant improvement of the average solution quality can be observed. Every alternative approach is outperformed. However, at the same time, the sensitivity of the actor-critic method increases such that a much higher number of additional iterations is required in order to gain improvements.

In summary, the computational study of Chap. 8 shows that it is possible to successfully realize approximate anticipation for a dynamic vehicle routing problem with late customer requests. The proposed approach leads to high quality solutions for each of the problem instances considered and outperforms the competing non-reactive approaches in by far most of the cases. Further improvements may be gained by considering alternative members of the family of actor-critic methods and by spending some effort on parameter tuning.

The good performance achieved by approximate anticipation for dynamic routing of a service vehicle is a strong aspect in favor of its application in real opeations. Nevertheless, the managerial impact of anticipatory optimization will be different within each opeational context. Chapter 9 outlines the general determinants of the impact and points out that in a business context, a degree of anticipation must be selected subject to a number of criteria. On the one hand the available technological resources in terms of infrastructure and competences must be considered. On the other hand, an adaptation of the degree of anticipation requires careful assessment of both the currently implemented way of decision making and the structure of the operations under consideration.

Summing up, the major contributions of the preceding chapters are

- definition of anticipatory optimization by means of degrees of anticipation,
- a concise illustration of the major approaches to perfect anticipation,
- integration of optimization and data mining into a unified methodology,
- identification of the synergies of optimization and data mining,
- classification of approaches from the literature according to the synergies of optimization and data mining,
- formulation of the generalized actor-critic framework in terms of optimization and data mining,
- classification of approaches to dynamic vehicle routing according to their degree of anticipation,
- a detailed case study of dynamic routing of a service vehicle including the determination of an optimal solution for the case,
- introduction of a new family of actor-critic methods for dynamic routing of a service vehicle,
- introduction of new approaches to non-reactive anticipation for dynamic routing of a service vehicle,
- the first successful realization of approximate anticipation for dynamic vehicle routing with late requests.

The opportunities for future research are manifold. On a general level both the generalized actor-critic framework and the synergies of optimization and data mining may be transferred to new operational contexts. Moreover, integrating new types of information structures into the actor-critic framework represents a challenge for future research.

Concerning the field of dynamic vehicle routing, the next steps to go comprise the realization of approximate anticipation for extended problem types. Consider dynamic routing of a fleet of service vehicles as an example. In addition, many operational contexts may require taking into account service times, time windows or vehicle capacities, to name only a few of the possible additional attributes.

Concerning dynamic routing of a service vehicle, an extensive analysis of the proposed family of actor-critic methods is needed. A detailed investigation promises both an improvement of the performance achieved so far and a better understanding of the problem structure.

References

Aarts E, Korst J (1989) Simulated Annealing and Boltzmann Machines. John Wiley & Sons, Chichester, UK

Abbiw-Jackson R, Golden B, Raghavan S, Wasil E (2006) A divide–and–conquer local search heuristic for data visualization. Comput Oper Res 33(11):3070–3087

Agafonov E, Bargiela A, Burke E, Peytchev E (2009) Mathematical justification of a heuristic for statistical correlation of real-life time series. Eur J Oper Res 198(1):275–286

Agrawal R, Srikant R (1994) Fast algorithms for mining association rules. In: Bocca J, Jarke M, Zaniolo C (eds) Proceedings of the 20th International Conference on Very Large Data Bases, Morgan Kaufmann, San Francisco, pp 487–499

Agrawal R, Imielinski T, Swami A (1993) Mining association rules between sets of items in large databases. In: Buneman P, Jajodia S (eds) Proceedings of the 1993 ACM SIGMOD International Conference on Management of Data, Washington D.C., ACM Press, New York, pp 207–216

Angelelli E, Bianchessi N, Mansini R, Speranza M (2009) Short term strategies for a dynamic multi-period routing problem. Transp Res C-Emerg 17(2):106–119

Angelelli E, Bianchessi N, Mansini R, Speranza M (2010) Comparison of policies in dynamic routing problems. J Oper Res Soc 61(4):686–695

Aoki M (1965) Optimal control of partially observable Markovian systems. J Frankl Inst 280(5):367–386

Applegate D, Bixby R, Chvatal V, Cook W (2007) The Traveling Salesman Problem. Princeton University Press, Princeton, NJ

Archetti C, Bertazzi L, Speranza M (2003) Reoptimizing the traveling salesman problem. Networks 42(3):154–159

Astrom K (1965) Optimal control of Markov processes with incomplete state information. J Math Anal Appl 10(1):174–205

Attanasio A, Bregman J, Ghiani G, Manni E (2007) Real-time fleet management at ecourier ltd. In: Zeimpekis V, Tarantilis C, Giaglis G, Minis I (eds) Dynamic Fleet Management, Springer, New York, pp 219–238

Ausiello G, Feuerstein E, Leonardi S, Stougie L, Talamo M (2001) Algorithms for the on-line traveling salesman. Algorithmica 29(4):560–581

Bains S (2003) Intelligence as physical computation. Interdiscip J Artif Intell Simul Behav 1(3):225–240

Bargiela A, Pedrycz W (2008) Towards a theory of granular computing for human-centred information processing. IEEE Trans Fuzzy Syst 16(2):320–330

Barto A, Bradtke S, Singh S (1993) Learning to act using real-time dynamic programming. Tech. Rep. 93-02, University of Massachussetts, Amherst, MA

Basu A (1998) Perspectives on operations research in data and knowledge management. Eur J Oper Res 111(1):1–14

Baykasoğlu A, Özbakir L (2007) Mepar-miner: Multi-expression programming for classification rule mining. Eur J Oper Res 183(2):767–784

Beale E (1988) Introduction to Optimization. John Wiley & Sons, Chichester, UK

Belacel N, Raval H, Punnen A (2007) Learning multicriteria fuzzy classification method PROFTAN from data. Comput Oper Res 34(7):1885–1898

Beliakov G, King M (2006) Density based fuzzy c-means clustering of non-convex patterns. Eur J Oper Res 173(3):717–728

Bellman R (1954) The theory of dynamic programming. Bull Am Math Soc 60(6):503–516

Bellman R (1957a) Dynamic Programming. Princeton University Press, Princeton, NJ

Bellman R (1957b) A Markovian decision porcess. J Math Mech 6(5):679–684

Bellman R, Dreyfus S (1959) Functional approximations in dynamic programming. Math Tables Other Aids Comput 13(68):247–251

Bent R, van Hentenryck P (2004) Scenario-based planning for partially dynamic vehicle routing with stochastic customers. Oper Res 52(6):977–987

Bent R, Van Hentenryck P (2007) Waiting and relocation strategies in online stochastic vehicle routing. In: Veloso M (ed) Proceedings of the Twentieth International Joint Conference on Artificial Intelligence, AAAI Press, Menlo Park, CA, pp 1816–1821

Bernataviciene J, Dzemyda G, Kurasova O, Marcinkevicius V (2006) Optimal decisions in combining the SOM with nonlinear projection methods. Eur J Oper Res 173(3):729–745

Berry M, Linoff G (2011) Data Mining Techniques, 3rd edn. Wiley, Indianapolis, IN

Bertazzi L, Bertsekas D, Speranza M (2001) Optimal and neuro-dynamic programming solutions for a stochastic inventory transportation problem. In: Kischka P, Leopold-Wildburger U, Mohring R, Radermacher F (eds) Models, Methods and Decision Support for Management, Physica-Verlag, Heidelberg, Germany, pp 65–78

Bertsekas D (1999) Nonlinear Programming, 2nd edn. Athena Scientific, Belmont, MA

Bertsekas D (2007) Dynamic Programming and Optimal Control, vol 2, 3rd edn. Athena Scientific, Belmont, MA

Bertsekas D, Tsitsiklis J (1989) Parallel and Distributed Computation: Numerical Methods. Prentice-Hall, Englewood Cliffs, NJ

Bertsekas D, Tsitsiklis J (1996) Neuro-Dynamic Programming. Athena Scientific, Belmont, MA

Bertsimas D, Howell L (1993) Further results on the probabilistic traveling salesman problem. Eur J Oper Res 65(1):68–95

Bertsimas D, van Ryzin G (1991) A stochastic and dynamic vehicle routing problem in the Euclidean plane. Oper Res 39(4):601–615

Bertsimas D, van Ryzin G (1993) Stochastic and dynamic vehicle routing problem in the Euclidean plane with multiple capacitated vehicles. Oper Res 41(1):60–76

Bianchi L (2000) Notes on dynamic vehicle routing. Tech. Rep. IDSIA-05-01, IDSIA (Istituto Dalle Molle di Studi sull'Intelligenza Artificiale), Manno-Lugano, Switzerland

Blum J (1954) Approximation methods which converge with probability one. Ann Math Stat 25(2):382–386

Boginski V, Butenko S, Pardalos P (2006) Mining market data: A network approach. Comput Oper Res 33(11):3171–3184

Bradley P, Fayyad U, Mangasarian O (1999) Mathematical programming for data mining: Formulations and challenges. INFORMS J Comput 11(3):217–238

Branke J, Middendorf M, Noeth G, Dessouky M (2005) Waiting strategies for dynamic vehicle routing. Transp Sci 39(3):298–312

Bräysy O, Gendreau M (2005) Vehicle routing problem with time windows. Transp Sci 39(1):104–118

Breimann L, Friedman J, Olshen R, Stone C (1984) Classification and Regression Trees. Chapman and Hall/CRC, New York

Brijs T, Swinnen G, Vanhoof K, Wets G (2004) Building an association rules framework to improve product assortment decisions. Data Min Knowl Discov 8(1):7–23

Bull L (ed) (2004) Applications of Learning Classifier Systems. Springer, Berlin, Germany

Burges C (1998) A tutorial on support vector machines for pattern recognition. Data Min Knowl Discov 2(2):121–167

Burkard R, Deineko V, van Dal R, van der Veen J, Woeginger G (1998) Well-solvable special cases of the traveling salesman problem: A survey. SIAM Rev 40(3):496–546

Burke E, Kendall G (2005) Search Methodologies. Springer, New York

Butz M (2002) Anticipatory Learning Classifier Systems. Springer, Berlin, Germany

Butz M, Lanzi P (2009) Sequential problems that test generalization in learning classifier systems. Evol Intell 2(3):141–147

Butz M, Sigaud O, Gerard P (2003) Anticipatory behavior: Exploiting knowledge about the future to improve current behavior. In: Butz M, Sigaud O, Gerard P (eds) Anticipatory Behavior in Adaptive Learning Systems: Foundations, Theories and Systems, Springer, Berlin, Germany, pp 1–10

Campbell D, Erdahl R, Johnson D, Bibelnieks E, Haydock M, Bullock M, Crowder H (2001) Optimizing customer mail streams at fingerhut. Interfaces 31(1):77–90

Carrizosa E, Martin-Barragan B (2006) Two-group classification via a biobjective margin maximization model. Eur J Oper Res 173(3):746–761

Cassandra A (1998) A survey of POMDP applications. Tech. rep., AAAI Fall Symposium on Planning with Partially Observable Markov Decision Processes, Orlando, FL

Chen M, Wu H (2005) An association-based clustering approach to order batching considering customer demand patterns. Omega 33(4):333–343

Chen M, Huang C, Chen K, Wu H (2005) Aggregation of orders in distribution centers using data mining. Expert Syst Appl 28(3):453–460

Chen V (2010) Preface to special volume on data mining. Ann Oper Res 174:1–2

Chen Z, Xu H (2006) Dynamic column generation for dynamic vehicle routing with time windows. Transp Sci 40(1):74–88

Cheung B, Choy K, Li C, Shi W, Tang J (2008) Dynamic routing model and solution methods for fleet management with mobile technologies. Int J Prod Econ 113(2):694–705

Chong E, Zak S (2008) An Introduction to Optimization, 3rd edn. John Wiley & Sons, Hoboken, NJ

Clarke G, Wright J (1964) Scheduling of vehicles from a central depot to a number of delivery points. Oper Res 12(4):568–581

Clemen R, Reilly T (2001) Making Hard Decisions. Duxbury, Belmont, CA

Cooper L, Giuffrida G (2000) Turning datamining into a management science tool: New algorithms and empirical results. Manag Sci 46(2):249–264

Crone S, Lessmann S, Stahlbock R (2006) The impact of preprocessing on data mining: An evaluation of classifier sensitivity in direct marketing. Eur J Oper Res 173(3):781–800

Dantzig G, Ramser J (1959) The truck dispatching problem. Manag Sci 6(1):80–91

Darken C, Moody J (1991) Towards faster stochastic gradient search. In: Moody J, Hanson D, Lippmann R (eds) Advances in Neural Information Processing Systems, Morgan Kauffman, San Mateo, CA, pp 1009–1016

Dayan P, Sejnowski T (1994) TD(λ) converges with probability 1. Mach Learn 14(3):295–301

Deisenroth M, Rasmussen C, Peters J (2009) Gaussian process dynamic programming. Neurocomputing 72(7-9):1508–1524

Delesie L, Croes L (2000) Operations research and knowledge discovery: A data mining method applied to health care management. Int Transact Oper Res 7(2):159–170

Denton F (1985) Data mining as an industry. Rev Econ Stat 67(1):124–127

Dror M (1993) Modeling vehicle routing with uncertain demands as a stochastic program: Properties of the corresponding solution. Eur J Oper Res 64(3):432–441

Dror M, Laporte G, Trudeau P (1989) Vehicle routing with stochastic demands: properties and solution frameworks. Transp Sci 23(3):166–176

Dupacova J, Sladky K (2002) Comparison of multistage stochastic programs with recourse and stochastic dynamic programs with discrete time. ZAMM – J Appl Math Mech 82(11-12): 753–765

Ehmke J, Meisel S (2008a) Charakterisierung des städtischen Straßenverkehrs mit Floating Car Data und Data Mining. Straßenverkehrstechnik 10

Ehmke J, Meisel S (2008b) Ermittlung dynamischer Fahrzeiten für die City-Logistik. In: Mattfeld D, Voß S, Suhl L, Gnther H (eds) Informations- und Kommunikationssysteme in Supply Chain Management, DSOR Lab, Paderborn, Germany, pp 107–122

Ehmke J, Meisel S, Engelmann S, Mattfeld D (2009) Data chain management for planning in city logistics. Int J Data Min Modell Manag 1(4):335–356

Ehmke J, Meisel S, Mattfeld D (2010) Floating car data based analysis of urban travel times for the provision of traffic quality. In: Barcelo J, Kuwahara M (eds) Traffic Data Collection and its Standardization, Springer, New York, pp 107–122

Eksioglu B, Vural A, Reisman A (2009) The vehicle routing problem: A taxonomic review. Comp Ind Eng 57(44):1472–1483

Fayyad U, Piatetsky-Shapiro G, Smyth P (1996) From data mining to knowledge discovery: An overview. In: Fayyad U, Piatetsky-Shapiro G, Smyth P, Uthurusamy R (eds) Advances in Knowledge Discovery and Data Mining, MIT Press, Menlo Park, CA, pp 1–34

Feng J, Kusiak A (2006) Data mining applications in engineering design, manufacturing and logistics. Int J Prod Res 44(14):2689–2694

Feo T, Resende M (1989) A probabilistic heuristic for a computationally difficult set covering problem. Oper Res Lett 8(2):67–71

Feo T, Resende M (1995) Greedy randomized adaptive search procedures. J Glob Optim 6(2): 109–133

Festa P, Resende M (2001) GRASP: An annotated bibliography. Tech. rep., AT&T Labs Research, Florham Park, NJ

Fisher R (1936) Annals of eugenics. Manag Sci 7(2):179–188

Flood M (1956) The traveling-salesman problem. Oper Res 4(1):61–75

French S (1988) Decision Theory. John Wiley & Sons, Chichester, UK

Friedman J (1997) Data mining and statistics: What's the connection? In: Proceedings of the 29th Symposium on the Interface: Computing Science and Statistics, Houston, TX

Gendreau M, Potvin J (1998) Dynamic vehicle routing and dispatching. In: Crainic T, Laporte G (eds) Fleet Management and Logistics, Kluwer, London, pp 115–126

Gendreau M, Guertin F, Potvin JY, Taillard E (1999) Parallel tabu search for real-time vehicle routing and dispatching. Transp Sci 33(4):381–390

George A, Powell W (2006) Adaptive stepsizes for recursive estimation with applications in approximate dynamic programming. Mach Learn 65(1):167–198

Ghiani G, Guerriero F, Laporte G, Musmanno R (2003) Real-time vehicle routing: solution concepts, algorithms and parallel strategies. Eur J Oper Res 151(16):1–11

Ghiani G, Manni E, Quaranta A, Triki C (2009) Anticipatory algorithms for same-day courier dispatching. Transp Res E-Logist 45(1):96–106

Glover F, Laguna M (1997) Tabu Search. Kluwer, Boston

Glowacka K, Henry R, May J (2009) A hybrid data mining/simulation approach for modelling outpatient no-shows in clinic scheduling. J Oper Res Soc 60(8):1056–1068

Goel A (2009) Vehicle scheduling and routing with drivers' working hours. Transp Sci 43(1):17–26

Goldberg D (1989) Genetic Algorithms in Search, Optimization, and Machine Learning. Addison-Wesley, Boston

Golden B, Raghavan S, Wasil E (eds) (2008) The Vehicle Routing Problem: Latest Advances and New Challenges. Springer, New York

Gosavi A (2009) Reinforcement learning: A tutorial survey and recent advances. INFORMS J Comput 21(2):178–192

Greco S, Matarazzo B, Slowinski R (2002) Multicriteria classification by dominance-based rough set approach. In: Kloesgen W, Zytkow J (eds) Handbook of Data Mining and Knowledge Discovery, Oxford University Press, New York, pp 318–327

Greco S, Matarazzo B, Slowinski R (2005) Decision rule approach. In: Figueira J, Greco S, Ehrgott M (eds) Multiple Criteria Decision Analysis: State of the Art Surveys, Springer, New York, pp 507–562

Gurvits L, Lin L, Hanson S (1994) Incremental learning of evaluation functions for absorbing Markov chains: New methods and theorems. Tech. rep., Siemens Corporate Research, Princeton, NJ

Han J, Pei J, Yin Y, Mao R (2004) Mining frequent patterns without candidate generation: A frequent-pattern tree approach. Data Min Knowl Discov 8(1):53–87

Hand D (1998) Data mining: Statistics and more? Am Stat 52(2):112–118

Hand D (1999) Statistics and data mining: Intersecting disciplines. SIGKDD Explor Newsl 1(1):16–19

Hand D, Blunt G, Kelly M, Adams N (2000) Data mining for fun and profit. Stat Sci 15(2):111–126

Hand D, Mannila H, Padhraic S (2001) Principles of Data Mining, 1st edn. MIT Press, Cambridge, MA

Hansen P, Jaumard B, Bruno S (2002) Polynomial algorithms for nested univariate clustering. Discret Math 245(1):93–105

Hartl R, Hasle G, Janssens G (2006) Editorial: Special issue on rich vehicle routing. Central Eur J Oper Res 14(2):103–104

Holland J (1975) Adaptation in Natural and Artificial Systems, 1st edn. University of Michigan Press, Ann Arbor, MI

Howard R (1960) Dynamic Programming and Markov Processes. MIT Press, Cambridge, MA

Huyet A (2006) Optimization and analysis aid via data-mining for simulated production systems. Eur J Oper Res 173(3):827–838

Hvattum L, Lokketangen A (2007) A branch–and–regret heuristic for stochastic and dynamic vehicle routing problems. Networks 49(4):330–340

Hvattum L, Lokketangen A, Laporte G (2006) Solving a dynamic and stochastic vehicle routing problem with a sample scenario hedging heuristic. Transp Sci 40(4):421–438

Ichoua S, Gendreau M, Potvin JY (2000) Diversion issues in real-time vehicle dispatching. Transp Sci 34(4):426–438

Ichoua S, Gendreau M, Potvin JY (2006) Exploiting knowledge about future demands for real-time vehicle dispatching. Transp Sci 40(2):211–225

Innis T (2006) Seasonal clustering technique for time series data. Eur J Oper Res 175(1):376–384

Jaillet P (1985) Probabilistic traveling salesman problems. PhD thesis, Massachusetts Institute of Technology, Cambridge, MA

Jaillet P (1988) A priori solution of a traveling salesman problem in which a random subset of the customers are visited. Oper Res 36(6):929–936

Janssens D, Brijs T, Vanhoof K, Wets G (2006) Evaluating the performance of cost-based discretization versus entropy- and error-based discretization. Comput Oper Res 33(11):3107–3123

Jones D, Collins A, Hand C (2006) A classification model based on goal programming with non-standard preference functions with application to the prediction of cinema-going behaviour. Eur J Oper Res 177(1):515–524

Kaelbling L, Littman M, Moore A (1996) Reinforcement learning: A survey. J Artif Intell Res 4:237–285

Kall P, Wallace S (1995) Stochastic Programming. John Wiley & Sons, New York

Karasözen B, Rubinov A, Weber GW (2006) Optimization in data mining. Eur J Oper Res 173(3):701–704

Kesten H (1958) Accelerated stochastic approximation. Ann Math Stat 29(1):41–59

Kiechle G, Doerner K, Gendreau M, Hartl R (2008) Waiting strategies for regular and emergency patient transportation. In: Fleischmann B, Borgwardt K, Klein R, Tuma A (eds) Operations Research Proceedings 2008, Springer, Berlin, Germany, pp 271–276

Kilby P, Prosser P, Shaw P (1998) Dynamic VRPs: A study of scenarios. Tech. Rep. APES-06-1998, University of Strathclyde, Glasgow, UK

Kilby P, Prosser P, Shaw P (2000) A comparison of traditional and constraint-based heuristic methods on vehicle routing problems with side constraints. Constraints 5(4):389–414

Kim SH, Nelson B (2001) A fully sequential procedure for indifference-zone selection in simulation. ACM Transact Modell Comput Simul 11(3):251–273

Kirkpatrick S, Gelatt C, Vecchi M (1983) Optimization by simulated annealing. Science 220(4598):671–680

Kleinberg J, Papadimitriou C, Raghavan P (1998) A microeconomic view of data mining. Data Min Knowl Discov 2(4):311–324

Kohonen T (1995) Self-Organizing Maps, 3rd edn. Springer, Berlin, Germany

Kulkarni G, Fathi Y (2007) Integer programming models for the q-mode problem. Eur J Oper Res 182(2):612–625

Larsen A, Madsen O, Solomon M (2002) Partially dynamic vehicle routing – models and algorithms. J Oper Res Soc 53(6):637–646

Larsen A, Madsen O, Solomon M (2004) The a priori dynamic traveling salesman problem with time windows. Transp Sci 38(4):459–472

Larsen A, Madsen O, Solomon M (2007) Classification of dynamic vehicle routing systems. In: Zeimpekis V, Tarantilis C, Giaglis G, Minis I (eds) Dynamic Fleet Management, Springer, New York, pp 19–40

Larsen A, Madsen O, Solomon M (2008) Recent developments in dynamic vehicle routing systems. In: Golden B, Raghavan S, Wasil E (eds) The Vehicle Routing Problem: Latest Advances and New Challenges, Springer, New York, pp 199–218

Leigh W, Hightower R, Modani N (2005) Forecasting the New York Stock Exchange composite index with past price and interest rate on condition of volume spike. Expert Syst Appl 28(1):1–8

Li L, Tang H, Wu Z, Gong J, Gruidl M, Zou J, Tockman M, Clark R (2004) Data mining techniques for cancer detection using serum proteomic profiling. Artif Intell Med 32(2):71–83

Li ST, Shue LY, Lee SF (2006) Enabling customer relationship management in ISP services through mining usage patterns. Expert Syst Appl 30(3):621–632

Li X, Olafsson S (2005) Discovering dispatching rules using data mining. J Sched 8(6):515–527

Li XB, Jacob V (2008) Adaptive data reduction for large-scale transaction data. Eur J Oper Res 188(3):910–924

Lin R (2009) Potential use of FP-growth algorithm for identifying competitive suppliers in SCM. J Oper Res Soc 60(8):1135–1141

Longman (2003) Longman Dictionary of Contemporary English, 4th edn. Pearson, White Plains, NY

Loshin D (2003) Business Intelligence. Morgan Kaufmann, New York

Lovell M (1983) Data mining. Rev Econ Stat 65(1):1–12

MacQueen J (1967) Some methods for classification and analysis of multivariate observations. In: Proceedings of 5-th Berkeley Symposium on Mathematical Statistics and Probability, vol 1, University of California Press, Berkeley, CA, pp 281–297

Mangasarian O (1965) Linear and nonlinear separation of patterns by linear programming. Oper Res 13(3):444–452

Mangasarian O (1997) Mathematical programming in data mining. Data Min Knowl Discov 1(2):183–201

Maxwell M, Restrepo M, Henderson S, Topaloglu H (2010) Approximate dynamic programming for ambulance redeployment. INFORMS J Comput 22(2):266–281

McClain J (1974) Dynamics of exponential smoothing with trend and seasonal terms. Manag Sci 20(9):1300–1304

Meiri R, Zahavi J (2006) Using simulated annealing to optimize the feature selection problem in marketing applications. Eur J Oper Res 171(3):842–858

Meisel S, Mattfeld D (2007a) Integration of data mining and operations research. In: Wang J (ed) Encyclopedia of Data Warehousing and Mining, vol II, 2nd edn, Information Science Reference, Hershey, PA, pp 1046–1052

Meisel S, Mattfeld D (2007b) Synergies of data mining and operations research. In: Proceedings of the 40th Hawaii International Conference on System Sciences, IEEE Computer Society, Los Alamitos, CA, p 56c

Meisel S, Mattfeld D (2010) Synergies of operations research and data mining. Eur J Oper Res 206(1):1–10

Meisel S, Suppa U, Mattfeld D (2010) GRASP based approximate dynamic programming for dynamic routing of a vehicle. In: Casatera M, Voss S (eds) Post Conference Volume of the VIII Metaheuristic International Conference, Springer, Berlin, Germany

Meisel S, Suppa U, Mattfeld D (2011) Serving multiple urban areas with stochastic customer requests. In: Kreowski HJ, Scholz-Reiter B, Thoben KD (eds) Dynamics in Logistics, Springer, Berlin, Germany, pp 59–68

Merriam-Webster (2003) Merriam-Webster Collegiate Dictionary, 11th edn. Merriam Webster Dictionaries, Springfield, MA

Mesarovic M, Takahara Y (1989) Abstract Systems Theory. Springer, Berlin, Germany

Mitrovic-Minic S, Laporte G (2004) Waiting strategies for the dynamic pickup and delivery problem with time windows. Transp Res B-Methodol 38(7):635–655

Mitrovic-Minic S, Krishnamurti R, Laporte G (2004) Double-horizon based heuristics for the dynamic pickup and delivery problem with time windows. Transp Res B-Methodol 38(8):669–685

Nahmias S (2008) Production and Operation Analysis, 6th edn. McGraw-Hill, Boston

Nemhauser G, Rinnooy Kan A, Todd M (1989) Optimization. Elsevier Science, Amsterdam

Novoa C, Storer R (2009) An approximate dynamic programming approach for the vehicle routing problem with stochastic demands. Eur J Oper Res 196(2):509–515

Olafsson S (2006) Introduction to operations research and data mining. Comput Oper Res 33(11):3067–3069

Olafsson S, Li X, Wu S (2008) Operations research and data mining. Eur J Oper Res 187(3):1429–1448

Osei-Bryson KM, Rayward-Smith V (2009) Data mining and operational research: techniques and applications. J Oper Res Soc 60(8):1043–1044

Ötztürk A, Kayahgil S, Özdemirel N (2006) Manufacturing lead time estimation using data mining. Eur J Oper Res 173(2):683–700

Padmanabhan B, Tuzhilin A (2003) On the use of optimization for data mining: Theoretical interactions and eCRM opportunities. Manag Sci 49(10):1327–1343

Papastavrou J (1996) A stochastic and dynamic routing policy using branching processes with state dependent immigration. Eur J Oper Res 95(1):167–177

Pendharkar P (2006) A data mining-constraint satisfaction optimization problem for cost effective classification. Comput Oper Res 33(11):3124–3135

Peng J (1993) Efficient dynamic programming-based learning for control. PhD thesis, Northeastern University, Boston

Piramuthu S (1996) Feed-forward neural networks and feature construction with correlation information: an integrated framework. Eur J Oper Res 93(2):418–427

Powell W (2007) Approximate Dynamic Programming. John Wiley & Sons, Hoboken, NJ

Powell W, Jaillet P, Odoni A (1995) Stochastic and dynamic routing and networks. In: Ball M, Magnanti T, Monma C, Nemhauser G (eds) Network Routing, Elsevier Science, Amsterdam, pp 141–295

Powell W, Towns M, Marar A (2000) On the value of optimal myopic solutions for dynamic routing and scheduling problems in the presence of user noncompliance. Transp Sci 34(1):67–85

Powell W, Shapiro J, Simao H (2001) A representational paradigm for dynamic resource transformation problems. Ann Oper Res 104(1-4):231–279

Proper S, Tadepalli P (2006) Scaling model-based average-reward reinforcement learning for product delivery. In: Lecture Notes in Computer Science, vol 4212, Springer, Berlin, Germany, pp 766–777

Psaraftis H (1988) Dynamic vehicle routing problems. In: Golden B, Assad A (eds) Vehicle Routing: Methods and Studies, North-Holland, Amsterdam, pp 233–248

Psaraftis H (1995) Dynamic vehicle routing: Status and prospects. Ann Oper Res 61(1):143–164

Puterman M (2005) Markov Decision Processes. John Wiley & Sons, Hoboken, NJ

Puterman M, Shin M (1978) Modified policy iteration algorithms for discounted Markov decision problems. Manag Sci 24(11):1127–1137

Quinlan J (1986) Induction of decision trees. Mach Learn 1(1):81–106

Quinlan J (1993) C4.5: Programs for Machine Learning. Morgan Kaufmann, San Mateo, CA

Raheja D, Llinas J, Nagi R, Romanowski C (2006) Data fusion/data mining-based architecture for condition-based maintenance. Int J Prod Res 44(14):2869–2887

Rao M (1971) Cluster analysis and mathematical programming. J Am Stat Assoc 66(335):622–626

Reeves C (1995) Introduction. In: Reeves C (ed) Modern Heuristic Techniques For Combinatorial Problems, McGraw-Hill, London

Robbins H, Monro S (1951) A stochastic approximation method. Ann Math Stat 22(3):400–407

Romanowski C, Nagi R (2004) A data mining approach to forming generic bills of materials in support of variant design activities. J Comput Inf Sci Eng 4(4):316–328

Romanowski C, Nagi R, Sudit M (2006) Data mining in engineering design environment: OR applications from graph matching. Comput Oper Res 33(11):3150–3160

Rosen R (1985) Anticipatory Systems: Philosophical, Mathematical and Methodological Foundations. Pergamon Press, Oxford, UK

Savelsbergh M, Sol M (1995) The general pickup and delivery problem. Transp Sci 29(1):17–29

Savelsbergh M, Sol M (1998) DRIVE: Dynamic routing of independent vehicles. Oper Res 46(4):474–490

Sağlam B, Salman F, Saym S, Türkay M (2006) A mixed-integer programming approach to the clustering problem with an application in customer segmentation. Eur J Oper Res 173(3):866–879

Sawicki P, Zak J (2008) Technical diagnostic of a fleet of vehicles using rough sets theory. Eur J Oper Res 193(3):891–903

Schneeweiss C (2003) Distributed Decision Making, 2nd edn. Springer, Berlin, Germany

Secomandi N (2000) Comparing neuro-dynamic programming algorithms for the vehicle routing problem with stochastic demands. Comput Oper Res 27(11-12):1201–1225

Secomandi N (2001) A rollout policy for the vehicle routing problem with stochastic demands. Oper Res 49(5):796–802

Secomandi N (2003) Analysis of a rollout approach to sequencing problems with stochastic routing applications. J Heuristics 9(4):321–352

Secomandi N, Margot F (2009) Reoptimization approaches for the vehicle-routing problem with stochastic demands. Oper Res 57(1):214–230

Séguin R, Potvin JY, Gendreau M, Crainic T, Marcotte P (1997) Real-time decision problems: an operational research perspective. J Oper Res Soc 48(2):162–174

Shao X, Wang Z, Li P, Feng C (2006) Integrating data mining and rough set for customer group-based discovery of product configuration rules. Int J Prod Res 44(14):2789–2811

Shearer C (2000) The CRISP-DM model: The new blueprint for data mining. J Data Warehous 5(4):13–22

Shen C, Chen Y (2008) A dynamic-programming algorithm for hierarchical discretization of continuous attributes. Eur J Oper Res 184(2):636–651

Shi L, Olafsson S (2000) Nested partitions method for global optimization. Oper Res 48(3):390–407

Si J, Barto A, Powell W, Wunsch D (2004) Handbook of Learning and Approximate Dynamic Programming. IEEE Press, New York

Sigaud O, Wilson S (2007) Learning classifier systems: A survey. Soft Comput 11(11):1065–1078

Simao H, Day J, George A, Gifford T, Nienow J, Powell W (2009) An approximate dynamic programming algorithm for large-scale fleet management: A case application. Transp Sci 43(2):178–197

Singh S, Dayan P (1998) Analytical mean squared error curves in temporal difference learning. Mach Learn 32(1):5–40

Slowinski R, Greco S, Matarazzo B (2005) Rough set based decision support. In: Burke E, Kendall G (eds) Search Methodologies: Introductory Tutorials in Optimization and Decision Support Techniques, Springer, New York, pp 475–527

Slowinski R, Greco S, Matarazzo B (2008) Rough sets in decision making. In: Meyers R (ed) Encyclopedia of Complexity and System Science, Springer, Berlin, Germany, pp 7753–7787

Smith K, Gupta J (2000) Neural networks in business: Techniques and applications for the operations researcher. Comput Oper Res 27(11-12):1023–1044

Solomon M (1987) Algorithms for the vehicle routing and scheduling problem with time windows. Oper Res 35(2):254–265
Spivey M, Powell W (2004) The dynamic assignment problem. Transp Sci 38(4):399–419
Stompe F, Meisel S (2006) Real-time data minig. Tech. rep., Carl-Friedrich Gauss Department, University of Braunschweig, Braunschweig, Germany
Sutton R, Barto A (1998) Reinforcement Learning. MIT Press, Cambridge, MA
Swihart M, Papastavrou J (1999) A stochastic and dynamic model for the single-vehicle pick-up and delivery problem. Eur J Oper Res 114(3):447–464
Thomas B (2007) Waiting strategies for anticipating service requests from known customer locations. Transp Sci 41(3):319–331
Thomas B, White III C (2004) Anticipatory route selection. Transp Sci 38(4):473–487
Thomas B, White III C (2007) The dynamic shortest path problem with anticipation. Eur J Oper Res 176(2):836–854
Topaloglu H, Powell W (2006) Dynamic-programming approximations for stochastic time-staged integer multicommodity-flow problems. INFORMS J Comput 18(1):31–42
Trafalis T, Gilbert R (2006) Robust classification and regression using support vector machines. Eur J Oper Res 173(3):893–909
Tseng J, Huang C, Jiang F, Ho J (2006) Applying a hybrid data-mining approach to prediction problems: A case of preferred suppliers prediction. Int J Prod Res 44(14):2935–2954
Tsitsiklis J (1994) Asynchronous stochastic approximation and Q-learning. Mach Learn 16(3):185–202
TSPLIB (2010) Library of sample instances for the TSP. URL http://www.iwr.uni-heidelberg.de/groups/comopt/software/TSPLIB95/index.html
Üney F, Türkay M (2006) A mixed-integer programming approach to multi-class data classification problem. Eur J Oper Res 173(3):910–920
Urbanowicz R, Moore J (2009) Learning classifier systems: A complete introduction, review, and roadmap. J Artif Evol Appl 2009, URL doi:10.1155/2009/736398
Van Hemert J, La Poutre J (2004) Dynamic routing problems with fruitful regions: Models and evolutionary computation. In: Yao X, Burke E, Lozano J, Smith J, Merelo Guervos J, Bullinaria J, Rowe J, Tino P, Kaban A, Schwefel HP (eds) Parallel Problem Solvers from Nature VIII, Springer, no. 3242 in Lecture Notes in Computer Science, pp 692–701
Van Hentenryck P, Bent R (2006) Online Stochastic Combinatorial Optimization. MIT-Press, Cambridge, MA
Van Roy B (2001) Neuro-dynamic programming: Overview and recent trends. In: Feinberg E, Shwarts A (eds) Handbook of Markov Decision Processes: Methods and Applications, Kluwer, Boston, pp 1–34
Watkins C (1989) Learning from delayed rewards. PhD thesis, Cambridge University, Cambridge, UK
Watkins C, Dayan P (1992) Q-learning. Mach Learn 8(3-4):279–292
White D, Sofge D (1992) Handbook of Intelligent Control. Von Nostrand Reinhold, New York
White J (1991) An existence theorem for OR/MS. Oper Res 39(2):183–193
Wu C (2006) Applying frequent itemset mining to identify a small itemset that satisfies a large percentage of orders in a warehouse. Comput Oper Res 33(11):3161–3170
Wu X, Yu P, Piatetsky-Shapiro G, Cercone N, Lin T, Kotagiri R, Wah B (2003) Data mining: How research meets practical development? Knowl Inform Syst 5(2):248–261
Yang J, Olafsson S (2005) Intelligent partitioning for feature selection. INFORMS J Comput 17(3):339–355
Yang J, Olafsson S (2006) Optimization-based feature selection with adaptive instance sampling. Comput Oper Res 33(11):3088–3106
Yang J, Olafsson S (2009) Near-optimal feature selection for large databases. J Oper Res Soc 60(8):1045–1055
Yuan F, Cheu R (2003) Incident detection using support vector machines. Transp Res C-Emerg 11(3-4):309–328

Zhang J, Shi Y, Zhang P (2008) Several multi-criteria programming methods for classification. Comput Oper Res 36(3):823–836

Index

Kolibri 360 GmbH
Ferdinand-Jühlke-Straße 7
99095 Erfurt, Deutschland
produktsicherheit@kolibri360.de